Bettina Berendt, Dunja Mladenič, Marco de Gemmis, Giovanni Semeraro,
Myra Spiliopoulou, Gerd Stumme, Vojtěch Svátek, and Filip Železný (Eds.)

Knowledge Discovery Enhanced with Semantic and Social Information

T0142068

Studies in Computational Intelligence, Volume 220

Editor-in-Chief
Prof. Janusz Kacprzyk
Systems Research Institute
Polish Academy of Sciences
ul. Newelska 6
01-447 Warsaw
Poland
E-mail: kacprzyk@ibspan.waw.pl

Bettina Berendt, Dunja Mladenič, Marco de Gemmis,
Giovanni Semeraro, Myra Spiliopoulou, Gerd Stumme,
Vojtěch Svátek, and Filip Železný (Eds.)

Knowledge Discovery Enhanced with Semantic and Social Information

Bettina Berendt
Katholieke Universiteit Leuven
Department of Computer Science
Celestijnenlaan 200A
3001 Heverlee
Belgium

E-mail: Bettina.Berendt@cs.kuleuven.be

Marco de Gemmis
Università degli Studi "Aldo Moro"
Dipartimento di Informatica
Via E. Orabona, 4
70126 - Bari
Italy

E-mail: degemmis@di.uniba.it

Dunja Mladenič
J. Stefan Institute
Jamova 39
1000 Ljubljana
Slovenia

E-mail: dunja.mladenic@ijs.si

Giovanni Semeraro
Università degli Studi "Aldo Moro"
Dipartimento di Informatica
Via E. Orabona, 4
70126 - Bari
Italy

E-mail: semeraro.di.uniba.it

Prof. Dr. Myra Spiliopoulou
Otto-von-Guericke-Universitaet Magdeburg
ITI/FIN, Faculty of Computer Science
P.O. Box 4120, D-39016 Magdeburg
Germany

E-mail: myra@iti.cs.uni-magdeburg.de

Prof. Dr. Gerd Stumme
Fachgebiet Wissensverarbeitung
Universität Kassel
Wilhelmshöher Allee 73
34121 Kassel
Germany

E-mail: stumme@cs.uni-kassel.de

Vojtěch Svátek
University of Economics, Prague
Nam.W.Churchilla 4
13067 Praha 3
Czech Republic

E-mail: Svatek@vse.cz

Filip Železný
Czech Technical University
Faculty of Electrical Engineering
Department of Cybernetics
Technicka 2
16627 Prague 6
Czech Republic

E-mail: zelezny@fel.cvut.cz

ISBN 978-3-642-42609-4   ISBN 978-3-642-01891-6 (eBook)

DOI 10.1007/978-3-642-01891-6

Studies in Computational Intelligence   ISSN 1860949X

© 2009 Springer-Verlag Berlin Heidelberg
Softcover re-print of the Hardcover 1st edition 2009

Typeset & Cover Design: Scientific Publishing Services Pvt. Ltd., Chennai, India.

Printed in acid-free paper

9 8 7 6 5 4 3 2 1

springer.com

Preface

There is general agreement that the quality of Machine Learning and Knowledge Discovery output strongly depends not only on the quality of source data and sophistication of learning algorithms, but also on additional, task/domain specific input provided by domain experts for the particular session. There is however less agreement on whether, when and how such input can and should effectively be formalized and reused as explicit prior knowledge.

In the first of the two parts into which the book is divided, we aimed to investigate current developments and new insights on learning techniques that exploit prior knowledge and on promising application areas. With respect to application areas, experiments on bio-informatics / medical and Web data environments are described. This part comprises a selection of extended contributions to the workshop *Prior Conceptual Knowledge in Machine Learning and Knowledge Discovery (PriCKL)*, held at ECML/PKDD 2007 18th European Conference on Machine Learning and 11th European Conference on Principles and Practice of Knowledge Discovery in Databases). The workshop is part of the activities of the "SEVENPRO – Semantic Virtual Engineering for Product Design" project of the European 6th Framework Programme.

The second part of the book has been motivated by the specification of Web 2.0. We observe Web 2.0 as a powerful means of promoting the Web as a social medium, stimulating interpersonal communication and fostering the sharing of content, information, semantics and knowledge among people. Chapters are authored by participants to the workshop *Web Mining 2.0*, held at ECML/PKDD 2007. The workshop hosted research on the role of web mining in and for the Web 2.0. It is part of the activities of the working groups "Ubiquitous Data – Interaction and Data Collection" and "Human Computer Interaction and Cognitive Modelling" of the Coordination Action "KDubiq – Knowledge Discovery in Ubiquitous Environments" of the European 6th Framework Programme.

The contributions to Part 1 addressed four different topics: inductive logic programming; the role of the human user; investigations of fully automated methods of integrating background knowledge; and the use of background knowledge for Web mining.

On Ontologies as Prior Conceptual Knowledge in Inductive Logic Programming by Francesca A. Lisi and Floriana Esposito provides an overview of Inductive Logic Programming attempts at using Ontologies as prior conceptual knowledge. Specifically, they compare the proposals CARIN-ALN and AL-log.

In *A Knowledge-Intensive Approach for Semi-Automatic Causal Subgroup Discovery*, Martin Atzmüller and Frank Puppe emphasize the role of the human expert in contributing background knowledge for data mining quality. They present a method for identifying causal relations between subgroups to form a network of links between subgroups. Background knowledge is used to add missing links in this network, correct directionality, and remove wrong links. Their approach is semi-automatic: the network and the relations are visualized to allow a user to accept them into a final causal model or not. An example case study illustrates how data mining can help to identify risk factors for medical conditions.

Fully automated uses of background knowledge and their advantages, including speed and accuracy, are investigated with respect to different types of machine learning in two contributions. *A study of the SEMINTEC approach to frequent pattern mining* by Joanna Józefowska, Agnieszka Lawrynowicz, and Tomasz Lukaszewski describes an experimental investigation of various settings under an approach to frequent pattern mining in description logics (DL) knowledge bases. Background knowledge is used to prune redundant (partial) patterns, which substantially speeds up pattern discovery. *Partitional Conceptual Clustering of Web Resources Annotated with Ontology Languages* by Floriana Esposito, Nicola Fanizzi, and Claudia dAmato proposes a way of clustering objects described in a logical language. The clustering method relies on a semi-distance measure and combines bisecting k-means and medoids into a hierarchical extension of the PAM algorithm (Partition Around Medoids).

Prior conceptual knowledge has a large importance for the Web. Approaches range from the use of background knowledge (or "semantics") to improve the results of mining Web resources, to the use of background knowledge in mining various other resources to improve the Web. Two contributions illustrate this broad range. *The Ex Project: Web Information Extraction using Extraction Ontologies* by Martin Labský, Vojtěch Svátek, Marek Nekvasil and Dušan Rak addresses the problem of using background knowledge for extracting knowledge from the Web. They use richly-structured extraction ontologies to aid the Information Extraction task. The system also makes it possible to re-use third-party ontologies and the results of inductive learning for subtasks where pre-labeled data abound. *Dealing with Background Knowledge in the SEWEBAR Project* by Jan Rauch and Milan Šimůnek illustrates the use of data mining with background knowledge for creating the Semantic Web. The goal is to generate, in a decentralized fashion, local analytical reports about a domain, and then to combine them, on the Semantic Web, into global analytical reports. Background knowledge is applied on a meta-level to guide the functioning of the mining algorithms themselves (in this case, GUHA). An example of background knowledge are useful category boundaries / granularity for quantitative attributes. The case study concerns relationships between health-related behavioral variables.

Part 2 accommodates the extended version of two contributions for the "Web Mining 2.0" workshop: The study of Linas Baltrunas and Francesco Ricci on *Item Weighting Techniques for Collaborative Filtering* deals with the dissemination of preferences about content. The authors investigate the use of item ratings for collaborative filtering in a recommendation engine. The authors study different methods for item weighting and item selection, with the intention to increase recommendation accuracy despite data sparsity and high dimensionality of the feature space.

In *Using Term-Matching Algorithms for the Annotation of Geo-Services*, Miha Grcar, Eva Klien, and Blaz Novak study semantic annotations of spatial objects: Their objective is to associate terms that describe the spatial objects with appropriate concepts from a domain ontology. Their method achieves this by associating terms with documents fetched from the Web and then assessing term similarity (and term/concept similarity) on the basis of (a) document similarity, (b) linguistic patterns and (c) Google distance.

We thank our reviewers for their careful help in selecting and improving submissions, the ECML/PKDD organizers and especially the Workshops Chairs for their support, our projects SEVENPRO and KDubiq, and the PASCAL project and the Czech Society for Cybernetics and Informatics for their financial support.

February 2009
Bettina Berendt
Marco de Gemmis
Dunja Mladenič
Giovanni Semeraro
Myra Spiliopoulou
Gerd Stumme
Vojtěch Svátek
Filip Železný

Contents

Part II: Web Mining 2.0

Part I
Prior Conceptual Knowledge in Machine Learning and Knowledge Discovery

On Ontologies as
Prior Conceptual Knowledge in
Inductive Logic Programming

Francesca A. Lisi and Floriana Esposito

Abstract. In this paper we consider the problem of having ontologies as prior conceptual knowledge in Inductive Logic Programming (ILP). In particular, we take a critical look at three ILP proposals based on knowledge representation frameworks that integrate Description Logics and Horn Clausal Logic. From the comparative analysis of the three, we draw general conclusions that can be considered as guidelines for an upcoming Onto-Relational Learning aimed at extending Relational Learning to account for ontologies.

1 Introduction

Exploiting domain-specific *prior knowledge* can significantly influence the output of Machine Learning (ML)/Knowledge Discovery (KD) algorithms. With the recent progress in both knowledge management technologies and ML/KD algorithms, many issues pertaining to how to exploit such knowledge have been opened. New challenges have also been raised by applications in areas requiring intensive accounting for background knowledge, such as bioinformatics and web mining. Among the different forms of knowledge that can be given as input to ML/KD algorithms, *conceptual knowledge* turns out to be more difficult to treat because it comes as a connected web of knowledge, a network in which the linking relationships are as prominent as the discrete bits of information. Conceptual knowledge includes most of commonsense knowledge and other encyclopedic types of knowledge such as *ontologies*. An ontology is a formal explicit specification of a shared conceptualization for a domain of interest [18]. Among the other things, this definition emphasizes the fact that an ontology has to be specified in a language that comes with

Francesca A. Lisi and Floriana Esposito
Dipartimento di Informatica, Università degli Studi di Bari
Via E. Orabona 4, 70125 Bari, Italy
e-mail: {lisi,esposito}@di.uniba.it

B. Berendt et al. (Eds.): Knowl. Disc. Enhan. with Sem. and Soc. Info., SCI 220, pp. 3–17.
springerlink.com   © Springer-Verlag Berlin Heidelberg 2009

a formal semantics. Only by using such a formal approach ontologies provide the machine interpretable meaning of concepts and relations that is expected when using an ontology-based approach. During the last decade increasing attention has been paid on ontologies and their role in Intelligent Systems as a means for conveying conceptual knowledge [7]. In Artificial Intelligence, an ontology refers to an engineering artifact (more precisely, produced according to the principles of *Ontological Engineering* [17]), constituted by a specific vocabulary used to describe a certain reality, plus a set of explicit assumptions regarding the intended meaning of the vocabulary words. This set of assumptions has usually the form of a First Order Logic (FOL) theory, where vocabulary words appear as unary or binary predicate names, respectively called concepts and relations. In the simplest case, an ontology describes a hierarchy of concepts related by subsumption relationships; in more sophisticated cases, suitable axioms are added in order to express other relationships between concepts and to constrain their intended interpretation. Among the formalisms proposed by Ontological Engineering, the most currently used are *Description Logics* (DLs) [1].

The use of *prior conceptual knowledge* is a core ingredient in Inductive Logic Programming (ILP) [36]. ILP was born at the intersection of Concept Learning [33] and Logic Programming [31]. Thus it has been historically concerned with rule induction from examples within the representation framework of Horn Clausal Logic (HCL) and with the aim of prediction. Currently ILP covers a broader area of research investigating - among other things - novel tasks like the descriptive ones that are more peculiar to Data Mining and novel approaches like the ones collectively known as Statistical Relational Learning. Though the use of background knowledge has been widely recognized as one of the strongest points of ILP when compared to other forms of Concept Learning, the background knowledge in ILP is often not organized around a well-formed conceptual model. This practice seems to ignore the growing demand for an ontological foundation of knowledge in intelligent systems. Rather, it highlights some difficulties in accommodating ontologies in ILP. Indeed the underlying Knowledge Representation (KR) frameworks (DLs and HCL, respectively) are deeply different in several respects but can be combined according to some limited forms of hybridization [41]. In this paper we take a critical look at three ILP proposals for these hybrid KR frameworks. We consider them as ILP attempts at having ontologies as prior conceptual knowledge.

The paper is organized as follows. Section 2 provides the essentials of DLs. Section 3 briefly describes different forms of integration of DLs and HCL. Section 4 first provides the basic notions of ILP, with a particular emphasis on the use of prior conceptual knowledge, then compares the three ILP proposals for hybrid DL-HCL KR formalisms. Section 5 concludes the paper with final remarks.

2 Ontologies and Description Logics

The most popular ontology languages are currently the ones based on Description Logics (DLs). DLs are a family of decidable FOL fragments that allow for the specification of knowledge in terms of classes (*concepts*), binary relations between classes (*roles*), and instances (*individuals*) [3]. Complex concepts can be defined from atomic concepts and roles by means of constructors (see Table 1). E.g., concept descriptions in the basic DL $\mathcal{AL}$ are formed according to only the constructors of atomic negation, concept conjunction, value restriction, and limited existential restriction. The DLs $\mathcal{ALC}$ and $\mathcal{ALN}$ are members of the $\mathcal{AL}$ family [44]. The former extends $\mathcal{AL}$ with (arbitrary) concept negation (also called complement and equivalent to having both concept union and full existential restriction), whereas the latter with number restriction. The DL $\mathcal{ALCNR}$ adds to the constructors inherited from $\mathcal{ALC}$ and $\mathcal{ALN}$ a further one: role intersection (see Table 1). Conversely, in the DL $\mathcal{SHIQ}$ [20] it is allowed to invert roles and to express qualified number restrictions of the form $\geq nS.C$ and $\leq nS.C$ where S is a simple role (see Table 1).

A DL knowledge base (KB) can state both is-a relations between concepts (*axioms*) and instance-of relations between individuals (resp. couples of individuals) and concepts (resp. roles) (*assertions*). Concepts and axioms form the

Table 1 Syntax and semantics of DLs

bottom (resp. top) concept	$\bot$ (resp. $\top$)	$\emptyset$ (resp. $\Delta^{\mathcal{I}}$)
atomic concept	A	$A^{\mathcal{I}} \subseteq \Delta^{\mathcal{I}}$
(abstract) role	R	$R^{\mathcal{I}} \subseteq \Delta^{\mathcal{I}} \times \Delta^{\mathcal{I}}$
(abstract) inverse role	R^-	$(R^{\mathcal{I}})^-$
(abstract) individual	a	$a^{\mathcal{I}} \in \Delta^{\mathcal{I}}$
concept negation	$\neg C$	$\Delta^{\mathcal{I}} \setminus C^{\mathcal{I}}$
concept intersection	$C_1 \sqcap C_2$	$C_1^{\mathcal{I}} \cap C_2^{\mathcal{I}}$
concept union	$C_1 \sqcup C_2$	$C_1^{\mathcal{I}} \cup C_2^{\mathcal{I}}$
value restriction	$\forall R.C$	$\{x \in \Delta^{\mathcal{I}} \mid \forall y\ (x,y) \in R^{\mathcal{I}} \to y \in C^{\mathcal{I}}\}$
existential restriction	$\exists R.C$	$\{x \in \Delta^{\mathcal{I}} \mid \exists y\ (x,y) \in R^{\mathcal{I}} \wedge y \in C^{\mathcal{I}}\}$
at least number restriction	$\geq nR$	$\{x \in \Delta^{\mathcal{I}} \mid |\{y|(x,y) \in R^{\mathcal{I}}\}| \geq n\}$
at most number restriction	$\leq nR$	$\{x \in \Delta^{\mathcal{I}} \mid |\{y|(x,y) \in R^{\mathcal{I}}\}| \leq n\}$
at least qualif. number restriction	$\geq nS.C$	$\{x \in \Delta^{\mathcal{I}} \mid |\{y \in C^{\mathcal{I}}|(x,y) \in S^{\mathcal{I}}\}| \geq n\}$
at most qualif. number restriction	$\leq nS.C$	$\{x \in \Delta^{\mathcal{I}} \mid |\{y \in C^{\mathcal{I}}|(x,y) \in S^{\mathcal{I}}\}| \leq n\}$
role intersection	$R_1 \sqcap R_2$	$R_1^{\mathcal{I}} \cap R_2^{\mathcal{I}}$
concept equivalence axiom	$C_1 \equiv C_2$	$C_1^{\mathcal{I}} = C_2^{\mathcal{I}}$
concept subsumption axiom	$C_1 \sqsubseteq C_2$	$C_1^{\mathcal{I}} \subseteq C_2^{\mathcal{I}}$
role equivalence axiom	$R \equiv S$	$R^{\mathcal{I}} = S^{\mathcal{I}}$
role inclusion axiom	$R \sqsubseteq S$	$R^{\mathcal{I}} \subseteq S^{\mathcal{I}}$
concept assertion	$a : C$	$a^{\mathcal{I}} \in C^{\mathcal{I}}$
role assertion	$\langle a, b \rangle : R$	$(a^{\mathcal{I}}, b^{\mathcal{I}}) \in R^{\mathcal{I}}$
individual equality assertion	$a \approx b$	$a^{\mathcal{I}} = b^{\mathcal{I}}$
individual inequality assertion	$a \not\approx b$	$a^{\mathcal{I}} \neq b^{\mathcal{I}}$

so-called TBox (Terminological Box, or intensional part of a DL KB) whereas individuals and assertions form the so-called ABox (Assertional Box, or extensional part of a DL KB). A $\mathcal{SHIQ}$ KB encompasses also a RBox (Role Box) which consists of axioms concerning abstract roles. The semantics of DLs is defined through a mapping to FOL [3]. An *interpretation* $\mathcal{I} = (\Delta^{\mathcal{I}}, \cdot^{\mathcal{I}})$ for a DL KB consists of a domain $\Delta^{\mathcal{I}}$ and a mapping function $\cdot^{\mathcal{I}}$. In particular, individuals are mapped to elements of $\Delta^{\mathcal{I}}$ such that $a^{\mathcal{I}} \neq b^{\mathcal{I}}$ if $a \neq b$ (*Unique Names Assumption* (UNA) [39]). Yet in $\mathcal{SHIQ}$ UNA does not hold by default [16]. Thus individual equality (inequality) assertions may appear in a $\mathcal{SHIQ}$ KB (see Table 1). Also the KB represents many different interpretations, i.e. all its models. This is coherent with the *Open World Assumption* (OWA) that holds in FOL semantics. The main reasoning task for a DL KB is the *consistency check* that is performed by applying decision procedures based on tableau calculus. Decidability of reasoning is crucial in DLs.

When a DL-based ontology language is adopted, an ontology is nothing else than a TBox. If the ontology is populated, it corresponds to a whole DL KB, i.e. encompassing also an ABox.

3   Ontologies and Logic Programming

3.1   Basics of Horn Clausal Logic

Logic Programming is rooted into a fragment of Clausal Logics (CLs) known as Horn Clausal Logic (HCL) [31]. The basic element in CLs is the *atom* of the form $p(t_i, \ldots, t_{k_i})$ such that each p is a predicate symbol and each t_j is a term. A *term* is either a constant or a variable or a more complex term obtained by applying a functor to simpler term. Constant, variable, functor and predicate symbols belong to mutually disjoint alphabets. A *literal* is an atom either negated or not. A *clause* is a universally quantified disjunction of literals. Usually the universal quantifiers are omitted to simplify notation. Alternative notations are a clause as set of literals and a clause as an implication. A *program* is a set of clauses. HCL admits only so-called definite clauses. A *definite clause* is an implication of the form

$$\alpha_0 \leftarrow \alpha_1, \ldots, \alpha_m$$

where $m \geq 0$ and α_i are atoms, i.e. a clause with exactly one positive literal. The right-hand side α_0 and the left-hand side $\alpha_1, \ldots, \alpha_m$ of the implication are called *head* and *body* of the clause, respectively. Note that the body is intended to be an existentially quantified conjunctive formula $\exists \alpha_1 \wedge \ldots \wedge \alpha_m$. Furthermore definite clauses with $m > 0$ and $m = 0$ are called *rules* and *facts* respectively. The *model-theoretic* **semantics** of HCL is based on the notion of Herbrand interpretation. The corresponding proof-theoretic semantics is based on the *Closed World Assumption* (CWA). Deductive **reasoning** with HCL is formalized in its proof theory. In clausal logic *resolution* comprises a single inference rule which, from any two clauses having an appropriate form, derives a new clause as their consequence. Resolution is sound: every resolvent is implied by its parents. It is also refutation complete: the empty clause is derivable by resolution from any set S of Horn clauses if S is unsatisfiable.

Definite clauses are also at the basis of deductive databases [6]. In particular, the language DATALOG for deductive databases does not allow functors and recursion. Based on the distinction between extensional and intensional predicates, a DATALOG program D can be divided into two parts, called extensional and intensional. The *extensional part*, denoted as $EDB(D)$, is the set of facts of D involving the extensional predicates, whereas the *intensional part* $IDB(D)$ is the set of all other clauses of D. The main reasoning task in DATALOG is query answering. A *query* Q to a DATALOG program D is a DATALOG clause of the form

$$\leftarrow \alpha_1, \ldots, \alpha_m$$

where $m > 0$, and α_i is a DATALOG atom. An *answer* to a query Q is a substitution θ for the variables of Q. An answer is correct with respect to the DATALOG program D if $D \models Q\theta$. The *answer set* to a query Q is the set of answers to Q that are correct w.r.t. D and such that $Q\theta$ is ground. In other words the answer set to a query Q is the set of all ground instances of Q which are logical consequences of D. Answers are computed by refutation.

Disjunctive DATALOG (DATALOG$^\vee$) is a variant of DATALOG where disjunctions may appear in the rule heads [11]. Therefore DATALOG$^\vee$ can not be considered a fragment of HCL. Advanced versions (DATALOG$^{\neg\vee}$) also allow for negation in the bodies, which can be handled according to a semantics for negation in CL. A DATALOG$^{\neg\vee}$ *program* Π is a set of DATALOG$^{\neg\vee}$ rules. If, for all $R \in \Pi$, $n \leq 1$, Π is called a DATALOG$^\neg$ program. If, for all $R \in \Pi$, $h = 0$, Π is called a positive DATALOG$^\vee$ program. If, for all $R \in \Pi$, $n \leq 1$ and $h = 0$, Π is called a positive DATALOG program. If there are no occurrences of variable symbols in Π, Π is called a *ground* program. Defining the semantics of a DATALOG$^{\neg\vee}$ program is complicated by the presence of disjunction in the rules' heads because it makes the underlying disjunctive logic programming inherently nonmonotonic, i.e. new information can invalidate previous conclusions. Among the many alternatives, one widely accepted semantics for DATALOG$^{\neg\vee}$ is the extension to the disjunctive case of the *stable model semantics* [15]. According to this semantics, a DATALOG$^{\neg\vee}$ program may have several alternative models (but possibly none), each corresponding to a possible view of the reality.

3.2 Combining DLs and HCL

The integration of ontologies into Logic Programming follows the tradition of KR research on *hybrid systems*, i.e. those systems which are constituted by two or more subsystems dealing with distinct portions of a single KB by performing specific reasoning procedures [14]. The motivation for investigating and developing such systems is to improve on two basic features of KR formalisms, namely *representational adequacy* and *deductive power*, by preserving the other crucial feature, i.e. *decidability*. Indeed DLs and CLs are FOL fragments incomparable as for the expressiveness [3] and the semantics [40][1] but combinable under certain conditions. In particular, combining DLs with CLs can easily yield to undecidability if the interface between the DL part and the CL part of the hybrid KB does not fulfill the condition of *safeness*, i.e. does not solve the semantic mismatch between DLs and HCL [34, 41].

$\mathcal{AL}$-**log** [10] is a hybrid KR system that integrates $\mathcal{ALC}$ and DATALOG. In particular, variables occurring in the body of rules (called *constrained* DATALOG *clauses*) may be constrained with $\mathcal{ALC}$ concept assertions to be used as 'typing constraints'. This makes rules applicable only to explicitly named objects. Reasoning for $\mathcal{AL}$-log knowledge bases is based on *constrained SLD-resolution*, i.e. an extension of SLD-resolution with a tableau calculus for $\mathcal{ALC}$ to deal with constraints. Constrained SLD-resolution is *decidable* and runs in single non-deterministic exponential time. Constrained SLD-refutation is a complete and sound method for answering *ground* queries.

A comprehensive study of the effects of combining DLs and HCL (more precisely, Horn rules) can be found in [23]. Here the family **Carin** of hybrid languages is presented. Special attention is devoted to the DL $\mathcal{ALCNR}$. The results of the study can be summarized as follows: (i) answering conjunctive queries over $\mathcal{ALCNR}$ TBoxes is decidable, (ii) query answering in a logic obtained by extending $\mathcal{ALCNR}$ with non-recursive DATALOG rules, where both concepts and roles can occur in rule bodies, is also decidable, as it can be reduced to computing a union of conjunctive query answers, (iii) if rules are recursive, query answering becomes undecidable, (iv) decidability can be regained by disallowing certain combinations of constructors in the logic, and (v) decidability can be regained by requiring rules to be *role-safe*, where at least one variable from each role literal must occur in some non-DL-atom. As in $\mathcal{AL}$-log, query answering is decided using constrained resolution and a modified version of tableau calculus.

The hybrid KR framework of $\mathcal{DL}$+**log** [42] allows for the tight integration of DLs and DATALOG$^{\neg\vee}$ [11]. More precisely, it allows a DL KB to be extended with *weakly-safe* DATALOG$^{\neg\vee}$ rules. The condition of weak safeness allows to overcome the main representational limits of the approaches based on the DL-safeness condition, e.g. the possibility of expressing conjunctive queries (CQ) and unions of conjunctive queries (UCQ)[2], by keeping the integration scheme still decidable. To a certain extent, $\mathcal{DL}$+log is between $\mathcal{AL}$-log and CARIN. For $\mathcal{DL}$+log two semantics have been defined: a first-order

---

[1] Remind that the OWA holds for DLs whereas CWA is valid in HCL. Note that the OWA and CWA have a strong influence on the results of reasoning.

[2] A *Boolean UCQ* over a predicate alphabet P is a first-order sentence of the form $\exists\mathbf{X}.conj_1(\mathbf{X}) \vee \ldots \vee conj_n(\mathbf{X})$, where $\mathbf{X}$ is a tuple of variable symbols and each $conj_i(\mathbf{X})$ is a set of atoms whose predicates are in P and whose arguments are either constants or variables from $\mathbf{X}$. A *Boolean CQ* corresponds to a Boolean UCQ in the case when $n = 1$.

logic (FOL) semantics and a nonmonotonic (NM) semantics. In particular, the latter extends the stable model semantics of DATALOG$^{\neg\vee}$ [15]. According to it, DL-predicates are still interpreted under OWA, while DATALOG predicates are interpreted under CWA. Notice that, under both semantics, entailment can be reduced to satisfiability and, analogously, that CQ answering can be reduced to satisfiability. The problem statement of satisfiability for finite $\mathcal{DL}$+log KBs relies on the problem known as the *Boolean CQ/UCQ containment problem*[3] in DLs. It is shown that the decidability of reasoning in $\mathcal{DL}$+log, thus of ground query answering, depends on the decidability of the Boolean CQ/UCQ containment problem in $\mathcal{DL}$. Currently, $\mathcal{SHIQ}$+log is the most expressive decidable instantiation of $\mathcal{DL}$+log.

The distinguishing features of these three hybrid DL-CL KR frameworks are summarized in Table 2.

Table 2 Three hybrid DL-CL KR frameworks

	Carin [23]	$\mathcal{AL}$-log [10]	$\mathcal{DL}$+log [42]
DL language	any DL	$\mathcal{ALC}$	any DL
CL language	Horn clauses	DATALOG clauses	DATALOG$^{\neg\vee}$ clauses
integration	unsafe	safe	weakly-safe
rule head literals	DL/Horn literals	DATALOG literal	DL/DATALOG literals
rule body literals	DL/Horn literals	$\mathcal{ALC}$/DATALOG literals (no roles)	DL/DATALOG$^{\neg}$ literals
semantics	Herbrand models+DL models	idem	stable models+DL models
reasoning	SLD-resolution+tableau calculus	idem	stable model computation + Boolean CQ/UCQ containment
decidability	only for some instantiations	yes	for all instantiations with DLs for which the Boolean CQ/UCQ containment is decidable
implementation	n.a.(?)	n.a.(?)	n.a.

4 Ontologies and Inductive Logic Programming

ILP was born at the intersection between Logic Programming and Concept Learning. From Logic Programming it has borrowed the KR framework. From Concept Learning it has inherited the inferential mechanisms for induction, the most prominent of which is *generalization*.

4.1 Induction in ILP

In Concept Learning generalization is traditionally viewed as search through a partially ordered space of inductive hypotheses [33]. According to this vision, an inductive hypothesis is a clausal theory and the induction of a single clause requires (i) structuring, (ii) searching and (iii) bounding the space of clauses.

First we focus on (i) by clarifying the notion of *ordering* for clauses. An ordering allows for determining which one, between two clauses, is more general than the other. Actually quasi-orders are considered, therefore uncomparable pairs of clauses are admitted. One such ordering is θ-*subsumption* [37]:

---

[3] This problem was called *existential entailment* in [23].

Given two clauses C and D, we say that C θ-subsumes D if there exists a substitution θ, such that $C\theta \subseteq D$.

Once structured, the space of hypotheses can be searched (ii) by means of refinement operators. The definition of refinement operators presupposes the investigation of the properties of the various quasi-orders. In Shapiro's sense [45], a refinement operator is a function which computes a set of specializations of a clause. Specialization is suited for the direction of search in his approach. His kind of refinement operator has been therefore called a *downward* refinement operator in ILP. Dually, operators can be also defined to compute generalizations of clauses. These can be applied in a bottom-up search, so they have been named *upward* refinement operators. A good refinement operator should satisfy certain desirable properties [22]. We shall illustrate these properties for the case of downward refinement operators but analogous conditions are actually required to hold for upward refinement operators as well. Ideally, a downward refinement operator should compute only a *finite* set of specializations of each clause - otherwise it will be of limited practical use. This condition is called *local finiteness*. Furthermore, it should be *complete*: every specialization should be reachable by a finite number of applications of the operator. Finally, it is better only to compute *proper* specializations of a clause, for otherwise repeated application of the operator might get stuck in a sequence of equivalent clauses, without ever achieving any real specialization. Operators that satisfy all these conditions simultaneously are called *ideal*. It has been shown that ideal upward/downward refinement operators do not exist for both full and Horn clausal languages ordered by either subsumption or the stronger orders (e.g. implication).

In order to define a refinement operator for full clausal languages, it is necessary to drop one of the three properties of idealness. Since local finiteness and completeness are usually considered the most important among these properties, this means that locally finite and complete, but improper refinement operators can be defined for full clausal languages. On the other hand, in order to retain all the three properties of idealness, it seems that the only possibility is to restrict the search space. Hence, the definition of refinement operators is usually coupled with the specification of a declarative bias for bounding the space of clauses (iii). *Bias* concerns anything which constrains the search for theories [46]. Following [35] we will distinguish three kinds of bias: (a) *Language bias* that specifies constraints on the clauses in the search space; (b) *Search bias* that has to do with the way an ILP system searches its space of permitted clauses; (c) *Validation bias* that concerns the stopping criterion of the ILP system.

Induction with ILP generalizes from individual instances/observations in the presence of background knowledge, finding *valid hypotheses*. Validity depends on the underlying *setting*. At present, there exist several formalizations of induction in clausal logic that can be classified according to the following two orthogonal dimensions: the *scope of induction* (discrimination vs characterization) and the *representation of observations* (ground definite clauses vs ground unit clauses) [9]. The first dimension impacts the learning problem itself. *Discrimination* aims at inducing hypotheses with discriminant power as required in tasks such as classification where observations encompass both positive and negative examples. *Characteristization* is more suitable for finding regularities in a data set. This corresponds to learning from positive examples only. For a more complete discussion of differences between discrimination and characterization see [32]. The second dimension affects the notion of *coverage*, i.e. the condition under which a hypothesis explains an observation. In *learning from entailment*, hypotheses are clausal theories, observations are ground definite clauses, and a hypothesis covers an observation if the hypothesis logically entails the observation. In *learning from interpretations*, hypotheses are clausal theories, observations are Herbrand interpretations (ground unit clauses) and a hypothesis covers an observation if the observation is a model for the hypothesis. A deeper investigation of the two logical settings can be found in [8].

4.2 *Prior Conceptual Knowledge in ILP*

The Logic Programming framework provides a uniform representation means to ILP. As a matter of fact, the use of background knowledge (BK) fits very naturally within ILP. In particular, the induced theory and the BK are of the same form. They just derive from different sources: the induced theory is generated by the ILP system, while BK is provided by the user of the ILP system. Note that the BK in ILP conveys prior conceptual knowledge on a certain domain that can be fruitfully incorporated into generality orders. Indeed combining the examples with what we already know often allows for the construction of a more satisfactory theory that can be glanced from the examples by themselves. Given the usefulness of BK, orders have been proposed that reckon with it, e.g. Plotkin's *relative subsumption* [38] and Buntine's *generalized subsumption* [4]. Relative subsumption applies to arbitrary clauses and the BK may be an arbitrary finite set of clauses. Generalized subsumption only applies to definite clauses and the BK should be a definite program. Each of these orders is related to some form of deduction. It can be shown by using these two forms of deduction that generalized subsumption is a weaker quasi-order than relative subsumption. Also, it can be shown that both relative and generalized subsumption reduce to ordinary subsumption in case of non-tautologous clauses and empty BK.

Generalized subsumption is of major interest to this paper: Given two definite clauses C and D standardized apart and a definite program $\mathcal{K}$, we say that $C \succeq_{\mathcal{K}} D$ iff there exists a ground substitution θ for C such that (i) $head(C)\theta = head(D)\sigma$ and (ii) $\mathcal{K} \cup body(D)\sigma \models body(C)\theta$ where σ is a

Skolem substitution[4] for D with respect to $\{C\}\cup\mathcal{K}$. Generalized subsumption is also called *semantic generality* in contrast to θ-subsumption which is a purely syntactic generality. In the general case, generalized subsumption is undecidable and does not introduce a lattice on a set of clauses. Yet for DATALOG it is decidable.

4.3 Ontologies in ILP

Hybrid KR systems combining DLs and (fragments of) HCL have very recently attracted some attention in the ILP community. Three ILP frameworks have been proposed that adopt a hybrid DL-HCL representation for both hypotheses and background knowledge: [43] chooses CARIN-$\mathcal{ALN}$, [24] resorts to $\mathcal{AL}$-log, and [26] builds upon $\mathcal{SHIQ}$+log. A comparative analysis of the three is reported in Table 3. They can be considered as attempts at accommodating ontologies in ILP. Indeed, they can deal with $\mathcal{ALN}$, $\mathcal{ALC}$, and $\mathcal{SHIQ}$ ontologies respectively. We remind the reader that $\mathcal{ALN}$ and $\mathcal{ALC}$ are incomparable DLs.

Table 3 Comparison between three ILP frameworks that take ontologies into account

	Learning in Carin-$\mathcal{ALN}$ [43]	Learning in $\mathcal{AL}$-log [24]	Learning in $\mathcal{SHIQ}$+log [26]
prior knowledge	CARIN-$\mathcal{ALN}$ KB	$\mathcal{AL}$-log KB	$\mathcal{SHIQ}$+log KB
ontology lang.	$\mathcal{ALN}$	$\mathcal{ALC}$	$\mathcal{SHIQ}$
rule lang.	Horn clauses	DATALOG clauses	DATALOG clauses
hypothesis lang.	CARIN-$\mathcal{ALN}$ non-recursive rules	constrained DATALOG clauses	$\mathcal{SHIQ}$+log non-recursive rules
target predicate	Horn literal	DATALOG literal	$\mathcal{SHIQ}$/DATALOG literal
setting	interpretations	interpretations/entailment	interpretations
induction	predictive	predictive/descriptive	predictive/descriptive
generality order	extension of [4] to CARIN-$\mathcal{ALN}$	extension of [4] to $\mathcal{AL}$-log	extension of [4] to $\mathcal{SHIQ}$+log
coverage test	CARIN-$\mathcal{ALN}$ query answering	$\mathcal{AL}$-log query answering	$\mathcal{SHIQ}$+log query answering
ref. operators	n.a.	downward	downward/upward
implementation	n.a.	partial	n.a.
application	n.a.	yes	n.a.

4.3.1 Learning in Carin-$\mathcal{ALN}$

The framework proposed in [43] focuses on discriminant induction and adopts the ILP setting of learning from interpretations. Hypotheses are represented as CARIN-$\mathcal{ALN}$ non-recursive rules with a Horn literal in the head that plays the role of target concept. The coverage relation of hypotheses against examples adapts the usual one in learning from interpretations to the case of hybrid CARIN-$\mathcal{ALN}$ BK. The generality relation between two hypotheses is defined as an extension of generalized subsumption. Procedures for testing

---

[4] Let $\mathcal{B}$ be a clausal theory and H be a clause. Let $X_1,\ldots,X_n$ be all the variables appearing in H, and $a_1,\ldots,a_n$ be distinct constants (individuals) not appearing in $\mathcal{B}$ or H. Then the substitution $\{X_1/a_1,\ldots,X_n/a_n\}$ is called a *Skolem substitution* for H w.r.t. $\mathcal{B}$.

both the coverage relation and the generality relation are based on the existential entailment algorithm of CARIN. Following [43], Kietz studies the learnability of CARIN-$\mathcal{ALN}$, thus providing a pre-processing method which enables ILP systems to learn CARIN-$\mathcal{ALN}$ rules [21].

4.3.2 Learning in $\mathcal{AL}$-log

In [24], hypotheses are represented as constrained DATALOG clauses that are linked, connected (or range-restricted), and compliant with the bias of Object Identity (OI)[5]. As opposite to [43], this framework is general, meaning that it is valid whatever the scope of induction (description/prediction) is. Therefore the literal in the head of hypotheses represents a concept to be either discriminated from others (*discriminant induction*) or characterized (*characteristic induction*). The generality relation for one such hypothesis language is an adaptation of generalized subsumption [4], named $\mathcal{B}$-subsumption, to the $\mathcal{AL}$-log KR framework. It gives raise to a quasi-order and can be checked with a decidable procedure based on constrained SLD-resolution [28]. Coverage relations for both ILP settings of learning from interpretations and learning from entailment have been defined on the basis of query answering in $\mathcal{AL}$-log [25]. As opposite to [43], the framework has been implemented in an ILP system [30]. More precisely, an instantiation of it for the case of *characteristic induction from interpretations* has been considered. Indeed, the system supports a variant of a very popular data mining task - frequent pattern discovery - where rich prior conceptual knowledge is taken into account during the discovery process in order to find patterns at multiple levels of description granularity. The search through the space of patterns represented as unary conjunctive queries in $\mathcal{AL}$-log and organized according to $\mathcal{B}$-subsumption is performed by applying an ideal downward refinement operator [29].

4.3.3 Learning in $\mathcal{SHIQ}$+log

The ILP framework presented in [26] represents hypotheses as $\mathcal{SHIQ}$+log rules restricted to positive DATALOG and organizes them according to a generality ordering inspired by generalized subsumption. The resulting hypothesis space can be searched by means of refinement operators either top-down or bottom-up. Analogously to [24], this framework encompasses both scopes of induction but, differently from [24], it assumes the ILP setting of learning from interpretations only. Both the coverage relation and the generality relation boil down to query answering in $\mathcal{SHIQ}$+log, thus can be reformulated as satisfiability problems. Compared to [43] and [24], this framework shows an added value which can be summarized as follows. First, it relies on a more expressive DL, i.e. $\mathcal{SHIQ}$. Second, it allows for inducing definitions

---

[5] The OI bias can be considered as an extension of the UNA from the semantic level to the syntactic one of $\mathcal{AL}$-log. It can be the starting point for the definition of either an equational theory or a quasi-order for constrained DATALOG clauses.

for new DL concepts, i.e. rules with a $\mathcal{SHIQ}$ literal in the head. Third, it adopts a tighter form of integration between the DL and the CL part, i.e. the weakly-safe one.

5   Conclusions and Future Directions

In this paper, we have revised the ILP literature addressing the issues raised by having ontologies as prior conceptual knowledge. Considering that standard ontology languages are nowadays based on DLs, only three ILP works have been found that propose a solution to these issues [43, 24, 26]. They adopt a hybrid DL-HCL formalism as KR framework, namely CARIN-$\mathcal{ALN}$, $\mathcal{AL}$-log, and $\mathcal{SHIQ}$+log. Closely related to DL-HCL KR systems are the hybrid formalims arising from the study of many-sorted logics, where a FOL language is combined with a sort language which can be regarded as an elementary DL [12]. In this respect the study of a sorted downward refinement [13] can be also considered as a contribution to the topic of this paper.

From the comparative analysis of [43],[24] and [26], a common feature emerges: Both proposals resort to Buntine's generalized subsumption and extend it in a non-trivial way. This choice is due to the fact that, among the semantic generality orders in ILP, generalized subsumption applies only to definite clauses, therefore suits well the hypothesis language in all the three ILP frameworks. Following these guidelines, new ILP frameworks can be designed to deal with more or differently expressive hybrid DL-CL languages according to the DL chosen (e.g., learning CARIN-$\mathcal{ALCNR}$ rules), or the clausal language chosen (e.g., learning recursive CARIN rules), or the integration scheme (e.g., learning CARIN rules with DL-literals in the head). An important requirement will be the definition of a *semantic* generality relation for hypotheses to take into account the background knowledge. Of course, generalized subsumption may turn out to be not suitable for these upcoming ILP frameworks, e.g. for the case of learning disjunctive $\mathcal{DL}$+log rules. Each of these frameworks will contribute to a new extension of Relational Learning, that we call Onto-Relational Learning, to account for ontologies in a clear, well-founded and systematic way analogously to what has been done in Statistical Relational Learning. Note that matching Table 3 against Table 2 one may figure out what is the state-of-the-art and what are the directions of research on learning by having ontologies as prior conceptual knowledge. Also he/she can get suggestions on what is the most appropriate among these ILP frameworks for a certain intended application. Also it would be interesting to investigate how the nature of rules (i.e., the intended context of usage) may impact the learning process as for the scope of induction and other variables in the learning problem statement. E.g., the problem of learning $\mathcal{AL}$-log rules for classification purposes differ greatly from the apparently similar learning problem faced in [30].

Ontologies and DLs are also playing a relevant role in the definition of the *Semantic Web* [2]. Indeed, the design of the ontology language for the

Semantic Web, OWL[6], has been based on the very expressive DL $\mathcal{SHOIN}(\mathbf{D})$ [19]. Whereas OWL is already undergoing the second round of the standardization process at W3C, the debate around a unified language for *rules* is still ongoing. There are proposals trying to extend OWL with constructs inspired to HCL in order 'to build rules on top of ontologies'. Acquiring Semantic Web rules is a task that can be automated by applying ML/KD algorithms such as the ones that can be derived from the two ILP frameworks analysed in this paper and the future ones in the stream of Onto-Relational Learning. E.g., the adoption of $\mathcal{DL}$+log *full*, i.e. including nonmonotonic features like negation and disjunction, as language for hypotheses and background knowledge will strengthen the ability of the ILP frameworks to deal with incomplete knowledge. One such ability can turn out to be useful in the Semantic Web, and complementary to reasoning with uncertainty and under inconsistency.

Besides theoretical issues, most future work will have to be devoted to implementation and application. When moving to practice, issues like efficiency and scalability become of paramount importance, especially in the Semantic Web context. These concerns may drive the attention of ILP research towards less expressive hybrid KR frameworks in order to gain in tractability, e.g. instantiations of $\mathcal{DL}$+log with DL-Lite [5]. Applications can come out of some of the many use cases for Semantic Web rules specified by the RIF W3C Working Group. Considering the current trend to have rules within ontologies rather than on top of ontologies, it is worthwhile to explore the possibility of learning rules for ontology evolution according to the proof-of-concept scenario described in [27].

References

1. Baader, F., Calvanese, D., McGuinness, D., Nardi, D., Patel-Schneider, P. (eds.): The Description Logic Handbook: Theory, Implementation and Applications. Cambridge University Press, Cambridge (2003)
2. Berners-Lee, T., Hendler, J., Lassila, O.: The Semantic Web. Scientific American (May 2001)
3. Borgida, A.: On the relative expressiveness of description logics and predicate logics. Artificial Intelligence 82(1–2), 353–367 (1996)
4. Buntine, W.: Generalized subsumption and its application to induction and redundancy. Artificial Intelligence 36(2), 149–176 (1988)
5. Calvanese, D., Lenzerini, M., Rosati, R., Vetere, G.: Dl-lite: Practical reasoning for rich dls. In: Haarslev, V., Möller, R. (eds.) Proceedings of the 2004 International Workshop on Description Logics (DL 2004), CEUR Workshop Proceedings, vol. 104. CEUR-WS.org (2004)
6. Ceri, S., Gottlob, G., Tanca, L.: Logic Programming and Databases. Springer, Heidelberg (1990)
7. Chandrasekaran, B., Josephson, J., Benjamins, V.: What are ontologies, and why do we need them? IEEE Intelligent Systems 14(1), 20–26 (1999)

---

[6] http://www.w3.org/2004/OWL/

8. De Raedt, L.: Logical Settings for Concept-Learning. Artificial Intelligence 95(1), 187–201 (1997)
9. De Raedt, L., Dehaspe, L.: Clausal Discovery. Machine Learning 26(2–3), 99–146 (1997)
10. Donini, F., Lenzerini, M., Nardi, D., Schaerf, A.: $\mathcal{AL}$-log: Integrating Datalog and Description Logics. Journal of Intelligent Information Systems 10(3), 227–252 (1998)
11. Eiter, T., Gottlob, G., Mannila, H.: Disjunctive DATALOG. ACM Transactions on Database Systems 22(3), 364–418 (1997)
12. Frisch, A.: The substitutional framework for sorted deduction: Fundamental results on hybrid reasoning. Artificial Intelligence 49, 161–198 (1991)
13. Frisch, A.: Sorted downward refinement: Building background knowledge into a refinement operator for inductive logic programming. In: Džeroski, S., Flach, P.A. (eds.) ILP 1999. LNCS (LNAI), vol. 1634, pp. 104–115. Springer, Heidelberg (1999)
14. Frisch, A., Cohn, A.: Thoughts and afterthoughts on the 1988 workshop on principles of hybrid reasoning. AI Magazine 11(5), 84–87 (1991)
15. Gelfond, M., Lifschitz, V.: Classical negation in logic programs and disjunctive databases. New Generation Computing 9(3/4), 365–386 (1991)
16. Glimm, B., Horrocks, I., Lutz, C., Sattler, U.: Conjunctive query answering for the description logic $\mathcal{SHIQ}$. Journal of Artificial Intelligence Research 31, 151–198 (2008)
17. Gómez-Pérez, A., Fernández-López, M., Corcho, O.: Ontological Engineering. Springer, Heidelberg (2004)
18. Gruber, T.: A translation approach to portable ontology specifications. Knowledge Acquisition 5, 199–220 (1993)
19. Horrocks, I., Patel-Schneider, P., van Harmelen, F.: From $\mathcal{SHIQ}$ and RDF to OWL: The making of a web ontology language. Journal of Web Semantics 1(1), 7–26 (2003)
20. Horrocks, I., Sattler, U., Tobies, S.: Practical reasoning for very expressive description logics. Logic Journal of the IGPL 8(3), 239–263 (2000)
21. Kietz, J.: Learnability of description logic programs. In: Matwin, S., Sammut, C. (eds.) ILP 2002. LNCS (LNAI), vol. 2583, pp. 117–132. Springer, Heidelberg (2003)
22. van der Laag, P.: An analysis of refinement operators in inductive logic programming. Ph.D. Thesis, Erasmus University, Rotterdam, The Netherlands (1995)
23. Levy, A., Rousset, M.C.: Combining Horn rules and description logics in CARIN. Artificial Intelligence 104, 165–209 (1998)
24. Lisi, F.: Building Rules on Top of Ontologies for the Semantic Web with Inductive Logic Programming. Theory and Practice of Logic Programming 8(03), 271–300 (2008)
25. Lisi, F., Esposito, F.: Efficient Evaluation of Candidate Hypotheses in $\mathcal{AL}$-log. In: Camacho, R., King, R., Srinivasan, A. (eds.) ILP 2004. LNCS (LNAI), vol. 3194, pp. 216–233. Springer, Heidelberg (2004)
26. Lisi, F., Esposito, F.: Foundations of Onto-Relational Learning. In: Železný, F., Lavrač, N. (eds.) ILP 2008. LNCS (LNAI), vol. 5194, pp. 158–175. Springer, Heidelberg (2008)
27. Lisi, F., Esposito, F.: Learning $\mathcal{SHIQ}$+log Rules for Ontology Evolution. In: Gangemi, A., Keizer, J., Presutti, V., Stoermer, H. (eds.) Semantic Web Applications and Perspectives (SWAP 2008), CEUR Workshop Proceedings, vol. 426 (2008)

28. Lisi, F., Malerba, D.: Bridging the Gap between Horn Clausal Logic and Description Logics in Inductive Learning. In: Cappelli, A., Turini, F. (eds.) AI*IA 2003. LNCS (LNAI), vol. 2829, pp. 49–60. Springer, Heidelberg (2003)
29. Lisi, F., Malerba, D.: Ideal Refinement of Descriptions in $\mathcal{AL}$-log. In: Horváth, T., Yamamoto, A. (eds.) ILP 2003. LNCS (LNAI), vol. 2835, pp. 215–232. Springer, Heidelberg (2003)
30. Lisi, F., Malerba, D.: Inducing Multi-Level Association Rules from Multiple Relations. Machine Learning 55, 175–210 (2004)
31. Lloyd, J.: Foundations of Logic Programming, 2nd edn. Springer, Heidelberg (1987)
32. Michalski, R.: A theory and methodology of inductive learning. In: Michalski, R., Carbonell, J., Mitchell, T. (eds.) Machine Learning: an artificial intelligence approach, vol. I. Morgan Kaufmann, San Mateo (1983)
33. Mitchell, T.: Generalization as search. Artificial Intelligence 18, 203–226 (1982)
34. Motik, B., Sattler, U., Studer, R.: Query Answering for OWL-DL with Rules. In: McIlraith, S.A., Plexousakis, D., van Harmelen, F. (eds.) ISWC 2004. LNCS, vol. 3298, pp. 549–563. Springer, Heidelberg (2004)
35. Nédellec, C., Rouveirol, C., Adé, H., Bergadano, F., Tausend, B.: Declarative bias in ILP. In: Raedt, L.D. (ed.) Advances in Inductive Logic Programming, pp. 82–103. IOS Press, Amsterdam (1996)
36. Nienhuys-Cheng, S., de Wolf, R.: Foundations of Inductive Logic Programming. In: Nienhuys-Cheng, S.-H., de Wolf, R. (eds.) Foundations of Inductive Logic Programming. LNCS (LNAI), vol. 1228. Springer, Heidelberg (1997)
37. Plotkin, G.: A note on inductive generalization. Machine Intelligence 5, 153–163 (1970)
38. Plotkin, G.: A further note on inductive generalization. Machine Intelligence 6, 101–121 (1971)
39. Reiter, R.: Equality and domain closure in first order databases. Journal of ACM 27, 235–249 (1980)
40. Rosati, R.: On the decidability and complexity of integrating ontologies and rules. Journal of Web Semantics 3(1) (2005)
41. Rosati, R.: Semantic and computational advantages of the safe integration of ontologies and rules. In: Fages, F., Soliman, S. (eds.) PPSWR 2005. LNCS, vol. 3703, pp. 50–64. Springer, Heidelberg (2005)
42. Rosati, R.: $\mathcal{DL}$+log: Tight integration of description logics and disjunctive datalog. In: Doherty, P., Mylopoulos, J., Welty, C. (eds.) Proc. of Tenth International Conference on Principles of Knowledge Representation and Reasoning, pp. 68–78. AAAI Press, Menlo Park (2006)
43. Rouveirol, C., Ventos, V.: Towards Learning in CARIN-$\mathcal{ALN}$. In: Cussens, J., Frisch, A.M. (eds.) ILP 2000. LNCS (LNAI), vol. 1866, pp. 191–208. Springer, Heidelberg (2000)
44. Schmidt-Schauss, M., Smolka, G.: Attributive concept descriptions with complements. Artificial Intelligence 48(1), 1–26 (1991)
45. Shapiro, E.: Inductive inference of theories from facts. Technical Report 624, Dept. of Computer Science. Yale University (1981)
46. Utgoff, P., Mitchell, T.: Acquisition of appropriate bias for inductive concept learning. In: Proceedings of the 2nd National Conference on Artificial Intelligence, pp. 414–418. Morgan Kaufmann, Los Altos (1982)

A Knowledge-Intensive Approach for Semi-automatic Causal Subgroup Discovery

Martin Atzmueller and Frank Puppe

Abstract. This paper presents a methodological view on knowledge-intensive causal subgroup discovery implemented in a semi-automatic approach. We show how to identify causal relations between subgroups by generating an extended causal subgroup network utilizing background knowledge. Using the links within the network we can identify causal relations, but also relations that are potentially confounded and/or effect-modified by external (confounding) factors. In a semi-automatic approach, the network and the discovered relations are presented to the user as an intuitive visualization. The applicability and benefit of the presented technique is illustrated by examples from a case-study in the medical domain.

1 Introduction

Subgroup discovery (e.g., [17, 9, 10, 2]) is a powerful approach for exploratory and descriptive data mining to obtain an overview of the interesting dependencies between a specific target (dependent) variable and usually many explaining (independent) variables; for example, the risk of coronary heart disease (target variable) is significantly higher in the subgroup of smokers with a positive family history than in the general population.

When interpreting and applying the discovered relations, it is often necessary to consider the patterns in a causal context. However, considering an association with a causal interpretation can often lead to incorrect results, due to the basic tenet of statistical analysis that association does not imply causation (cf., [5]): A subgroup may not be causal for the target group, and thus can be suppressed by other causal groups and thus become irrelevant. Then, the causal subgroups are better suited for

Martin Atzmueller and Frank Puppe
University of Würzburg, Department of Computer Science VI, Am Hubland,
97074 Würzburg, Germany
e-mail: {atzmueller,puppe}@informatik.uni-wuerzburg.de

B. Berendt et al. (Eds.): Knowl. Disc. Enhan. with Sem. and Soc. Info., SCI 220, pp. 19–36.
springerlink.com　　　　　　　　　　　　　　　© Springer-Verlag Berlin Heidelberg 2009

characterizing the target concept, and already explain the relations captured by the suppressed subgroup. Furthermore, the estimated *effect*, i.e., the quality of the subgroup may be due to associations with other *confounding* factors that were not considered in the quality computation. For instance, a relatively high/low quality of a subgroup may only be due to other variables that are associated with the independent variables and are a direct cause of the (dependent) target variable. Then, it is necessary to identify potential confounders, and to measure or to control their influence concerning the subgroup and the target concept. Let us assume, for example, that ice cream consumption and murder rates are highly correlated. However, this does not necessarily mean that ice cream incites murder or that murder increases the demand for ice cream. Instead, both ice cream and murder rates might be joint effects of a common cause, namely, hot weather.

In this paper, we present an approach for the semi-automatic detection of (true) causal subgroups and potentially confounded and/or effect-modified relations. We apply known subgroup patterns in a knowledge-intensive process as background knowledge that can be incrementally refined: The applied patterns represent subgroups that are *acausal*, i.e., have no causes, and subgroup patterns that are known to be directly causally related to other (target) subgroups. Additionally, both types of patterns can be combined, for example, in the medical domain certain variables such as *Sex* have no causes, and it is known that they are causal risk factors for certain diseases.

Using the patterns contained in the background knowledge, and a set of subgroups for analysis, we can construct a causal net containing relations between the subgroups. In a semi-automatic process, this network can be interactively inspected and analyzed by the user: It directly provides a visualization of the causal relations between the subgroups, and also provides for a possible explanation of these. By traversing the relations in the network, we can then identify causal relations, potential confounding and/or effect-modification. The approach has been implemented as a plugin for the VIKAMINE [1] system.

The rest of the paper is organized as follows: First, we discuss the background of subgroup discovery, the concept of confounding, and basic constraint-based causal analysis methods in Section 2. After that we present the knowledge-intensive causal discovery approach for detecting causal and confounding/effect-modified relations in Section 3. Exemplary results of the application of the presented approach are given in Section 4 using data from a fielded system in the medical domain. Section 5 discusses the features, limitations, and experiences with the application of the presented approach. Finally, Section 6 concludes the paper with a summary of the presented work and points out interesting directions for future work.

2 Background

In this section, we first introduce the necessary notions concerning the used knowledge representation, before we define the setting for subgroup discovery. After that, we introduce the concept of confounding, criteria for its identification, and describe basic constraint-based techniques for causal subgroup analysis.

---

[1] http://vikamine.sourceforge.net

2.1 Basic Definitions

Let Ω_A be the set of all attributes. For each attribute $a \in \Omega_A$ a range $dom(a)$ of values is defined. The set $\mathcal{V}_A$ is assumed to be the (universal) set of attribute values of the form $(a = v)$, where $a \in \Omega_A$ is an attribute and $v \in dom(a)$ is an assignable value. We consider nominal attributes only so that numeric attributes need to be discretized accordingly.

Let CB be the case base (data set) containing all available cases (instances): A case $c \in CB$ is given by the n-tuple

$$c = ((a_1 = v_1), (a_2 = v_2), \ldots, (a_n = v_n))$$

of $n = |\Omega_A|$ attribute values, $v_i \in dom(a_i)$ for each a_i.

2.2 Subgroup Discovery

The main application areas of subgroup discovery (e.g., [17, 9, 10, 2]) are exploration and descriptive induction, to obtain an overview of the relations between a (dependent) target variable and a set of explaining (independent) variables. As in the *MIDOS* approach [17], we consider subgroups that have the most unusual (distributional) characteristics with respect to the concept of interest given by a binary target variable. Therefore, not necessarily complete relations but also partial relations, i.e., (small) subgroups with "interesting" characteristics can be sufficient.

Subgroup discovery mainly relies on the subgroup description language, the quality function, and the search strategy. Often, heuristic methods (e.g., [10]) but also efficient exhaustive algorithms (e.g., the SD-Map algorithm [2]) are applied. The description language specifies the individuals belonging to the subgroup. For a common single-relational propositional language a subgroup description can be defined as follows:

Definition 1 (Subgroup Description). *A subgroup description* $sd = \{e_1, e_2, \ldots, e_n\}$ *is defined by the conjunction of a set of selectors* $e_i = (a_i, V_i)$*: Each of these are selections on domains of attributes,* $a_i \in \Omega_A, V_i \subseteq dom(a_i)$*. We define* Ω_E *as the set of all selectors and* Ω_{sd} *as the set of all possible subgroup descriptions.*

A quality function measures the interestingness of the subgroup and is used to rank these. Typical quality criteria include the difference in the distribution of the target variable concerning the subgroup and the general population, and the subgroup size.

Definition 2 (Quality Function). *Given a particular target variable* $t \in \Omega_E$*, a quality function* $q : \Omega_{sd} \times \Omega_E \rightarrow R$ *is used in order to evaluate a subgroup description* $sd \in \Omega_{sd}$*, and to rank the discovered subgroups during search.*

Several quality functions were proposed (cf., [17, 9, 10, 2]), e.g., the functions q_{BT} and q_{RG}:

$$q_{BT} = \frac{(p - p_0) \cdot \sqrt{n}}{\sqrt{p_0 \cdot (1 - p_0)}} \cdot \sqrt{\frac{N}{N - n}}, \quad q_{RG} = \frac{p - p_0}{p_0 \cdot (1 - p_0)}, n \geq \mathcal{T}_{Supp},$$

where p is the relative frequency of the target variable in the subgroup, p_0 is the relative frequency of the target variable in the total population, $N = |CB|$ is the size of the total population, and n denotes the size of the subgroup.

In contrast to the quality function q_{BT} (the classic binomial test), the quality function q_{RG} only compares the target shares of the subgroup and the total population measuring the *relative gain*. Therefore, a support threshold $\mathcal{T}_{Supp}$ is necessary to discover significant subgroups.

The result of subgroup discovery is a set of subgroups. Since subgroup discovery methods are not necessarily covering algorithms, the discovered subgroups can overlap significantly, and their estimated quality (effect) might be confounded by external variables. In order to reduce the redundancy of the subgroups and to identify potential confounding factors, methods for causal analysis can then be applied.

2.3   The Concept of Confounding

Confounding can be described as a bias in the estimation of the effect of the subgroup on the target concept due to attributes affecting the target concept that are not contained in the subgroup description [11, 12]. Thus, confounding is caused by a lack of comparability between the subgroup the and complementary group due to a difference in the distribution of the target concept caused by other factors.

2.3.1   Simpson's Paradox

An extreme case for confounding is presented by *Simpson's Paradox* [15]: The (positive) effect (association) between a given variable X and a variable T is countered by a negative association given a third factor F, i.e., X and T are negatively correlated in the subpopulations defined by the values of F. For binary variables X, T, F this can be formulated as

$$P(T|X) > P(T|\neg X), P(T|X,F) < P(T|\neg X,F), P(T|X,\neg F) < P(T|\neg X,\neg F),$$

i.e., the event X increases the probability of T in a given population while it decreases the probability of T in the subpopulations restricted by F and $\neg F$.

As an example, let us assume that there is a positive correlation between the event X that describes *people that do not consume soft drinks* and T specifying the diagnosis *diabetes* (see Figure 1).

This association implies that people not consuming soft drinks are affected more often by diabetes (50% non-soft-drinkers vs. 40% soft-drinkers). However, this is due to age, if older people (given by F) consume soft drinks less often than younger people, and if diabetes occurs more often for older people, inverting the effect (see Figure 2).

Combined	T	¬T	$\sum$	Rate (T)
X	25	25	50	50%
¬X	20	30	50	40%
$\sum$	45	55	100	

Fig. 1 Example: Simpson's Paradox – aggregate counts

Restricted on F	T	¬T	$\sum$	Rate (T)
X	24	16	40	60%
¬X	8	2	10	80%
$\sum$	32	18	50	

Restricted on ¬F	T	¬T	$\sum$	Rate (T)
X	1	9	10	10%
¬X	12	28	40	30%
$\sum$	13	37	50	

Fig. 2 Example: Simpson's Paradox – grouped counts

2.3.2 Criteria for Identifying Confounders

There are three criteria that can be used to identify a **confounding factor** F [11], given the factors X contained in a subgroup description and a target concept T:

1. A confounding factor F must be a *cause* for the target concept T, e.g., an independent risk factor for a certain disease.
2. The factor F must be associated/correlated with the set of subgroup (factors) X, i.e.,, there needs to be a (statistically significant) association between F and X.
3. A confounding factor F must *not* be (causally) affected by the subgroup (factors) X, i.e., there must not be a causal dependency between X and F.

However, these criteria are only necessary but not sufficient to identify confounders. If purely automatic methods are applied for detecting confounding, then such approaches may label some variables as confounders incorrectly, e.g., if the real confounders have not been measured, or if their contributions cancel out.

Thus, user interaction is rather important for validating confounded relations. Furthermore, the identification of confounding requires causal (background) knowledge since confounding is itself a causal concept [12]. Such background knowledge can be formalized in an interactive approach as discussed below.

2.3.3 Proxy Factors and Effect Modification

There are two phenomena that are closely related to confounding. First, a factor may only be associated with the subgroup but may be the real cause for the target concept. Then, the subgroup is only a **proxy factor**. Another situation is given by **effect modification**: Then, a third factor F does not necessarily need to be associated with the subgroup described by the factors X; F can be an additional factor that increases the effect of X in a certain subpopulation only, pointing to new subgroup descriptions that are interesting by themselves.

2.4   Constraint-Based Methods for Causal Subgroup Analysis

In general, the philosophical concept of causality refers to the set of all particular 'cause-and-effect' or 'causal' relations. A subgroup is causal for the target group, if in an ideal experiment [5] the probability of an object not belonging to the subgroup to be a member of the target group increases or decreases when the characteristics of the object are changed such that the object becomes a member of the subgroup.

For example, the probability that a patient survives (target group) increases if the patient received a special treatment (subgroup). Then, a redundant subgroup, that is, for example, conditionally independent from the target group given another subgroup, can be suppressed. Then the relation can already be completely explained by the conditioning subgroup and the independent subgroup does not provide any significant insight.

For causal analysis the subgroups are represented by binary variables that are true for an object (case) if it is contained in the subgroup, and false otherwise. For constructing a causal subgroup network, constraint-based methods are particularly suitable because of scalability reasons (cf., [5, 14]) since they only depend on simple statistical (in-)dependence tests, e.g., the χ^2-test for independence.

However, constraint-based methods make several assumptions (cf., [5]) with respect to the data and the correctness of the statistical tests. The crucial condition is the *Markov condition* depending on the assumption that the data can be expressed by a Bayesian network: Let X be a node in a causal Bayesian network, and let Y be any node that is not a descendant of X in the causal network. Then, the Markov condition holds if X and Y are independent conditioned on the parents of X.

The CCC and CCU rules [14] described below constrain the possible causal models by applying simple statistical tests: For subgroups s_1, s_2, s_3 represented by binary variables the χ^2-test for independence is utilized for testing their independence $ID(s_1, s_2)$, dependence $D(s_1, s_2)$ and conditional independence $CondID(s_1, s_2|s_3)$, as shown below. For the tests user-selectable thresholds are applied, e.g., $\mathscr{T}_1 = 1, \mathscr{T}_2 = 3.84$, or higher, respectively:

$$ID(s_1, s_2) \longleftrightarrow \chi^2(s_1, s_2) < \mathscr{T}_1, \qquad D(s_1, s_2) \longleftrightarrow \chi^2(s_1, s_2) > \mathscr{T}_2,$$
$$CondID(s_1, s_2|s_3) \longleftrightarrow \chi^2(s_1, s_2|s_3 = 0) + \chi^2(s_1, s_2|s_3 = 1) < 2 \cdot \mathscr{T}_1$$

Thus, the decision of (conditional) (in-)dependence is threshold-based, which is a problem causing potential errors if very many tests are performed (e.g., [16]). Therefore, we propose a semi-automatic approach featuring interactive analysis of the inferred relations. The CCC rule requires three pairwise dependent (*correlated*) variables while the CCU rule requires two dependence relations and one independence relation.

Definition 3 (CCC Rule). *Let X, Y, Z denote three variables that are pairwise dependent, i.e., $D(X, Y), D(X, Z), D(Y, Z)$; let X and Z become independent when conditioned on Y. In the absence of hidden and confounding variables we may infer that one of the following causal relations exists between X, Y and Z: $X \rightarrow Y \rightarrow Z$,*

$X \leftarrow Y \rightarrow Z,\quad X \leftarrow Y \leftarrow Z$. *However, if X has no causes, then the first relation is the only one possible, even in the presence of hidden and confounding variables.*

Definition 4 (CCU Rule). *Let X,Y,Z denote three variables: X and Y are dependent $(D(X,Y))$, Y and Z are dependent $(D(Y,Z))$, X and Z are independent $(ID(X,Z))$, but X and Z become dependent when conditioned on Y $(CondD(X,Z|Y))$. In the absence of hidden and confounding variables, we may infer that X and Z cause Y.*

3 Semi-automatic Causal Subgroup Discovery and Analysis

The approach for knowledge-intensive causal subgroup discovery is embedded in an incremental semi-automatic process. In the following, we first introduce the process model for causal subgroup discovery and analysis. After that, we present the necessary background knowledge for effective causal analysis. Next, we describe a method for constructing an extended causal subgroup network, and discuss how to identify confounded and effect-modified relations. After that, we discuss visualizations for supporting the analysis.

3.1 Process Model for Causal Subgroup Discovery and Analysis

The process model includes a discovery, (causal) analysis, evaluation and validation, and finally a knowledge extraction and formalization step. The individual steps of the process for semi-automatic causal subgroup discovery are shown in Figure 3 and discussed below.

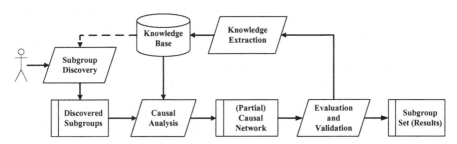

Fig. 3 Process Model for Knowledge-Intensive Causal Subgroup Analysis

1. First, the user applies a standard *subgroup discovery* approach, e.g., [10, 2]. The result of this step is a set of the most interesting subgroups. Optionally, background knowledge contained in the knowledge base can also be applied during subgroup discovery (e.g., as discussed in [3]).
2. Next, *causal analysis* using appropriate background knowledge for a detailed analysis of the discovered associations is applied. Using constraint-based

techniques for causal analysis, a (partial) causal subgroup network containing the discovered interesting subgroups is constructed.

3. In the *evaluation and validation* phase, the user assesses the (partial) causal network: Since the relations contained in the network can be wrong due to various statistical errors, inspection and validation of the causal relations is essential in order to obtain valid results. After inspection and validation the final results (as a selection of the discovered subgroup patterns) are obtained.

4. The user can extend and/or tune the applied background knowledge during the *knowledge extraction* step: Then, the knowledge base can be updated in incremental fashion by including further background knowledge, e.g., based upon the discovery results. The formalized background knowledge can then be incrementally (re-)applied and integrated into the process.

3.2   Extended Causal Subgroup Analysis

Detecting causal relations, i.e., (true) causal associations, confounding, and effect modification using causal subgroup analysis consists of the following two main steps that can be iteratively applied.

1. First, we generate a causal subgroup network considering a target group T, a user-selected set of subgroups U, a set of confirmed (causal) and potentially confounding factors C for any included group, a set of unconfirmed potentially confounding factors P given by subgroups significantly dependent with the target group, and additional background knowledge described below. In addition to causal links, the generated network also contains (undirected) associations between the variables.

2. In the next step, we traverse the network and mark the potential confounded and/or effect-modified relations. The causal network and the proposed relations are then presented to the user for subsequent interpretation and analysis. After the factors have been confirmed the background knowledge can be extended.

In the following sections we discuss these steps in detail. First, we describe the elements of the background knowledge that are utilized for causal analysis.

3.3   Background Knowledge for Causal Analysis

For the causal analysis step we first need to generate an extended causal network capturing the relations between the subgroups represented by binary variables.

In order to effectively generate the network, we need to include background knowledge provided by the user, for example, by a domain specialist. In the medical domain, for example, a lot of background knowledge is already available and can be directly integrated in the analysis phase. Examples include (causal) relations between diseases, and the relations between diseases and their respective findings. Additionally, it is often required to include the available background knowledge since otherwise only known relations are rediscovered (cf., [3]), valid results can even not be obtained, or too many results are proposed that need to be validated manually. The applicable background consists of two basic elements, i.e., acausal factors (variables) and causal relations between variables:

1. Acausal factors: These include factors represented by subgroups that have no causes; in the medical domain, e.g., the subgroup *Age* ≥ *70* or the subgroup *Sex* = *male* have no causes, while the subgroup *BMI* = *underweight* has certain causes.

2. (Direct) causal relations: Such relations include subgroups that are not only dependent but (directly) causal for the target group/target variable and/or other subgroups. In the medical domain, for example, the subgroup *Body-Mass-Index (BMI)=overweight* is directly causal for the subgroup *Gallstones=probable*.

Depending on the domain, for example, considering the medical domain, it is often relatively easy to provide acausal information. Direct and indirect causal relations are often also easy to acquire, and can be acquired 'on-the-fly' when applying the presented process. However, in some domains, e.g., in the medical domains, it is often difficult to provide non-ambiguous directed relationships between certain variables: One disease can cause another disease and vice versa, under different circumstances. In such cases, the relations should be formalized with respect to both directions, and can still be exploited in the method discussed below. Then, the interpretation performed by the user is crucial in order to obtain valid and ultimately interesting results.

3.4 Constructing an Extended Causal Subgroup Network

Algorithm 1. summarizes how to construct an extended causal subgroup network, based on a technique for basic causal subgroup analysis described in [9, 8]. When applying the algorithm, the relations contained in the network can be wrong due to various statistical errors (cf., [5]), especially for the CCU rule (cf., [14]). Therefore, after applying the algorithm, the resulting causal net is presented to the user for interactive analysis.

The first step (lines 1-5) of the algorithm determines for each subgroup pair (including the target group) whether they are independent, based on the inductive principle that the dependence of subgroups is necessary for their causality.

In the next step (lines 6-10) we determine for any pair of subgroups whether the first subgroup s_1 is suppressed by a second subgroup s_2, i.e., if s_1 is conditionally independent from the target group T given s_2. The χ^2-measure for the target group and s_1 is calculated both for the restriction on s_2 and its complementary subgroup. If the sum of the two test-values is below a threshold, then we can conclude that subgroup s_1 is conditionally independent from the target group. Conditional independence is a sufficient criterion, since the target distribution of s_1 can be explained by the target distribution in s_2, i.e., by the intersection. Since similar subgroups could symmetrically suppress each other, the subgroups are ordered by quality; then subgroups with a nearly identical extension (and lower quality) can be eliminated.

Algorithm 1. Constructing a causal subgroup net

Require: Target group T, user-selected set of subgroups U, potentially confounding groups P, background knowledge B containing acausal subgroup information, and known subgroup patterns $C \subseteq B$ that are directly causal for other subgroups. Define $\mathscr{S} = U \cup P \cup C$

1: **for all** $s_i, s_j \in \mathscr{S} \cup T, s_i \neq s_j$ **do**
2:    **if** $approxEqual(s_i, s_j)$ **then**
3:       Exclude any causalities for the subgroup with smaller correlation to T
4:    **if** $ID(s_i, s_j)$ **then**
5:       Exclude causality: $ex(s_i, s_j) = true$
6: **for all** $s_i, s_j \in \mathscr{S}, s_i \neq s_j$ **do**
7:    **if** $\neg ex(s_i, T)$, $\neg ex(s_j, T)$, or $\neg ex(s_i, s_j)$ **then**
8:       **if** $CondID(s_i, T | s_j)$ **then**
9:          Exclude causality: $ex(s_i, T) = true$, and include s_j into $separators(s_i, T)$
10:          If conditional independencies are symmetric, then select the strongest relation
11: **for all** $s_i, s_j \in \mathscr{S}, i < j$ **do**
12:    **if** $\neg ex(s_i, T)$, $\neg ex(s_j, T)$, or $\neg ex(s_i, s_j)$ **then**
13:       **if** $CondID(s_i, s_j | T)$ **then**
14:          Exclude causality: $ex(s_i, s_j) = true$, and include T into $separators(s_i, s_j)$
15: **for all** $s_i, s_j, s_k \in \mathscr{S}, i < j, i \neq k, j \neq k$ **do**
16:    **if** $\neg ex(s_i, s_j)$, $\neg ex(s_j, s_k)$, or $\neg ex(s_i, s_k)$ **then**
17:       **if** $CondID(s_i, s_j | s_k)$ **then**
18:          Exclude causality: $ex(s_i, s_j) = true$, and include s_k into $separators(s_i, s_j)$
19: Integrate direct causal links that are not conditionally excluded considering the sets C and B
20: **for all** $s_i, s_j, s_k \in \mathscr{S}$ **do**
21:    Apply the extended CCU rule, using background knowledge
22: **for all** $s_i, s_j, s_k \in \mathscr{S}$ **do**
23:    Apply the extended CCC rule, using background knowledge
24: **for all** $s_i, s_j, s_k \in \mathscr{S} \cup \{T\}, i < j, i \neq k, j \neq k$ **do**
25:    **if** $\neg CondID(s_i, s_j | s_k)$ **then**
26:       Integrate association between dependent s_i and s_j that are not conditionally excluded

---

The next two steps (lines 11-18) check conditional independence between each pair of subgroups given the target group or a third subgroup, respectively. For each pair of conditionally independent groups, the separating (conditioning) group is noted. Then, this separator information is exploited in the next steps, i.e., independencies or conditional independencies for pairs of groups derived in the first steps are used to exclude any causal links between the groups. The conditioning steps (lines 6-18) can optionally be iterated in order to condition on combinations of variables (pairs, triples). However, the decisions taken further (in the CCU and CCC rules) may become statistically weaker justified due to smaller counts in the considered contingency tables (e.g., [5, 8]).

Direct causal links (line 19) are added based on background knowledge, i.e., given subgroup patterns that are causal for specific subgroups. In the last step (lines 24-26) we also add conditional associations for dependent subgroups that are not conditionally independent and thus not suppressed by any other subgroups. Such links can later be useful in order to detect the (true) associations considering a confounding factor.

Extending CCC and CCU using Background Knowledge. The CCU and CCC steps (lines 20-23) derive the directions of the causal links between subgroups, based on information derived in the previous steps. In the context of the presented knowledge-intensive approach, we extend the basic CCC and CCU rules including background knowledge both for the derivation of additional links, and for inhibiting links that contradict the background knowledge. We introduce associations instead of causal directions if these are wrong, or if not enough information is available in order to derive the causal directions. The rationale behind this principle is given by the intuition that we want to exploit and provide as much information as possible considering the generated causal net. When identifying potentially confounded relations, we can also often utilize weaker associative information.

For the **extended** *CCU* **rule** we use background knowledge for inhibiting acausal directions, since the CCU rule can be disturbed by confounding and hidden variables. The causal or associative links do not necessarily indicate direct associations/causal links but can also point to relations enabled by hidden or confounding variables [14].

For the **extended** *CCC* **rule**, we can use the relations inferred by the extended CCU rule for disambiguating between the causal relations, if the *CCU* rule is applied in all possible ways: The non-separating condition (conditional dependency) of the relation identified by the *CCU* rule is not only a sufficient but a necessary condition [14], i.e., for $X \to Y \leftarrow Z$, with $CondID(X,Z|Y)$. Additionally, we can utilize background knowledge for distinguishing between the causal relations. So, for three variables X,Y,Z with $D(X,Y), D(X,Z), D(Y,Z)$, and $CondID(X,Z|Y)$, if there exists an (inferred) causal link $X \to Y$ between X and Y, we may identify the relation $X \to Y \to Z$ as the true relation. Otherwise, if Y or Z have no causes, then we select the respective relation, e.g., $X \leftarrow Y \to Z$ for an acausal variable Y.

3.5 Identifying Confounded Relations

A popular method for controlling confounding factors is given by *stratification* [11]: For example, in the medical domain a typical confounding factor is the attribute *age*: We can stratify on age groups such as $age < 30$, $age\ 30 - 69$, and $age \geq 70$. Then, the subgroup – target relations are measured within the different strata, and compared to the (crude) unstratified measure.

It is easy to see, that in the context of the presented approach stratification for a binary variable is equivalent to conditioning on them: If we assess a conditional subgroup – target relation and the subgroup factors become independent (or dependent), then this indicates potential confounding. After constructing a causal net, we can easily identify such relations. Since the causal directions derived by the extended CCC and CCU rules may be ambiguous, user interaction is crucial: In some domains, e.g., in the medical domains, it is often difficult to provide non-ambiguous directed relationships between certain variables: One disease can cause another disease and vice versa, under different circumstances. The network then also provides an intuitive visualization for the analysis.

In order to identify potentially confounded relations and the corresponding variables, as shown in Figure 4, and described below, we just need to traverse the network. In this way, we collect the interesting and relevant relations.

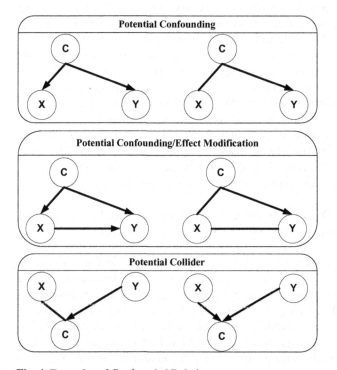

Fig. 4 Examples of Confounded Relations

- **Potential Confounding:** If there is an association between two variables X and Y, and the network contains the relations $C \to X$, $C \to Y$, and there is no causal link between X and Y, i.e., they are conditionally independent given C, then C is a confounder that inhibits the relation between X and Y. This is also true if there is no causal link between C and X but instead an association.
- **Potential Confounding/Effect Modification:** If the network contains the relations $C \to X$, $C \to Y$, and there is also either an association or a causal link between X and Y, then this points to confounding and possible effect modification of the relation between the variables X and Y.
- **Potential Collider (or Confounder):** If there is no (unconditioned) association between two variables X and Y and the network contains the relations $X \to C$

and $Y \rightarrow C$, then C is a potential collider: X and Y become dependent by conditioning on C. The variable C is then no confounder in the classical sense, if the (inferred) causal relations are indeed true. However, such a relation as inferred by the CCU rule can itself be distorted by confounded and hidden variables. The causal directions could also be inverted, if the association between X and Y is just not strong enough as estimated by the statistical tests. In this case, C is a potential confounder. Therefore, manual inspection is crucial in order to detect the true causal relation.

3.6 Supporting Visual Techniques

In order to support the user when analyzing and validating the causal network and the contained relations, we propose several techniques that are integrated into the main visualization for displaying the network itself. The proposed visualization is interactive such that the user can apply changes on the fly and adapt the parameters as needed. Feedback is then implemented instantly. The visualizations are completely integrated in the VIKAMINE (Visual, Interactive, and Knowledge-intensive Analysis and MINing Environment) system (vikamine.sourceforge.net) and can also be combined with the other visualization capabilites provided by the system. In this way, the causal analysis step is completely embedded within the proposed knowledge-intensive subgroup mining process. The approach for causal analysis is provided by a special plugin for the VIKAMINE system: Starting with the common subgroup discovery setting using the core functionality of VIKAMINE, the results can then be analyzed using the *causal analysis plugin* and can be visualized using the visualization techniques presented below.

In general, the *Visual Information Seeking Mantra* by [13], 'Overview first, zoom and filter, then details-on-demand' is an important guideline for visualization methods. In an iterative process, the user is able to subsequently concentrate on the interesting data by filtering irrelevant and redundant data, and focusing (zooming in) on the interesting elements, until finally details are available for an interesting subset of the analyzed objects. Therefore, we implemented the visualizations according to these guidelines. Therefore, we therefore propose the following techniques for the visualization of the causal subgroup network:

- **Association filters:** Selected associations can be explicitly excluded or included, e.g., with respect to non-causal associations. Then, these associations are marked as undirected lines otherwise. By default (non-causal) associations are excluded from view.
- **Color coding of nodes:** Acausal nodes, conflicting nodes, conflicting (bidirectional edges), are marked with different (customizable) colors for a better overview.
- **Node filters (zooming in on the network):** We provide filters for selecting a subset of the nodes of the network. The user can then select a set of specific nodes of interest in order to focus the analysis.

Fig. 5 Example for a causal
graph with filtered and
color-coded elements

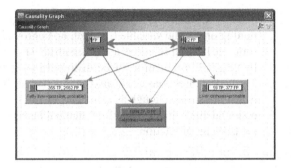

Essentially, the *assessment* of the patterns, i.e., the evaluation and validation of
the relations in order to determine their final interestingness need to be facilitated
by specialized techniques. Especially the zooming and filtering techniques allow an
easier assessment of the relations in the causal network as discussed below.

Figure 5 shows an example of a zoomed network for which several filters have
been applied. Compared to Figure 7 it can be easily seen that associations have been
filtered. The target node is shown in green color, while two obviously acausal edges
are colored in red, since incorrect causal directions have been derived for these (they
are only associated, as discussed above). Causal edges are shown in green, while the
conflicting bidirectional edges are shown in red. The strength of the association is
shown by the width of the edges. Also, in comparison to Figure 7 the subgroups
corresponding to the causal nodes are shown with their respective *true positive* and
false positive counts – as sub-bars within the nodes. In contrast, Figure 7 focuses on
the (causal) relations between the nodes.

4   Examples

For the experiments we utilized a case base containing about 8600 cases taken
from the SONOCONSULT system [7] – a medical documentation and consulta-
tion system for sonography. The system is in routine use in the DRK-hospital in
Berlin/Köpenick, and the collected cases contain detailed descriptions of findings
of the examination(s), together with the inferred diagnoses (binary attributes).

The experiments were performed using the VIKAMINE system [1] for semi-
automatic knowledge-intensive subgroup discovery, that was extended with a component implementing the presented approach.

In the following, we provide some (simplified) examples considering the diagnosis *Gallstones=established* as the target variable. After applying a subgroup discovery method, several subgroups were selected by the user in order to derive a causal
subgroup network, and to check the relations with respect to possible confounders.
These selected subgroups included, for example, the subgroup *Fatty liver=probable
or possible* and the subgroup *Liver cirrhosis=probable*.

A first result is shown in Figure 6: In this network, the subgroup *Liver cir-
rhosis=probable* is confounded by the variable *Age≥70*. However, there is still an

Fig. 6 Confounded Relation: Gallstones and Liver cirrhosis

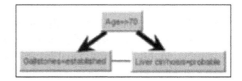

influence on the target variable considering the subgroup *Liver cirrhosis=probable* shown by the association between the subgroups. This first result indicates confounding and effect modification (the strengths of the association between the nodes is also visualized by the widths of the links). A more detailed result is shown in Figure 7: In this network another potential confounder, i.e., *Sex=female* is included. Then, it becomes clear, that both the subgroup *Fatty liver=probable or possible* and the subgroup *Liver cirrhosis=probable* are confounded by the variables *Sex* and *Age*, and the association (shown in Figure 6) between the subgroup *Liver cirrhosis=probable* and the target group is then no longer present (This example does not consider the removal of the gall-bladderwhich might have an additional effect concerning a medical interpretation).

It is easy to see that the generated causal subgroup network becomes harder to interpret, if many variables are included, and if the number of connections between the nodes increases. Therefore, the presented filtering and zooming techniques, e.g., in order to exclude (non-causal) associations, and for coloring the nodes and the edges of the network in order to increase their interpretability, are crucial for an effective implementation: Based on the available background knowledge, causal subgroup nodes, and (confirmed) causal directions can be marked. Since the network is first traversed and the potentially confounded relations are reported to the user, the analysis can also be focused towards the respective variables, as a further filtering condition. The user can then analyze selected parts of the network in more detail.

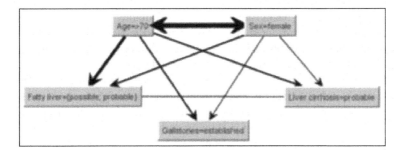

Fig. 7 Confounded Relations: Gallstones, Liver cirrhosis and Fatty Liver

5 Discussion

The presented approach has a rather broad application area in various domains. It provides the means for identifying confounded and/or effect-modified subgroups with respect to a certain target variable. Starting with the results (i.e., the output) of a common subgroup discovery algorithm, a causal network for a selected set of (interesting) subgroups is generated, that can then be traversed. Next, the interesting findings are highlighted for visual inspection by the user. In contrast to approaches that try to automatically detect confounding or try to locate instances of Simpson's Paradox, this is a quite important distinction: The presented method is not able to automatically detect confounded relations with guarantees for the correctness of these findings. Due to statistical errors of the applied tests, various assumptions for these, and the necessary causality conditions this is also a rather hard problem. However, the presented method allows for a semi-automatic approach for detection of confounding and/or effect-modification. This semi-automatic approach provides more reliable results, since the 'human in the loop' still has the option for the necessary validation of the identified relations, because the causal inferences made by the algorithm may be wrong due to statistical errors. This is especially relevant in the medical domain in which the findings of such an algorithm always need to be validated before being proclaimed as authorative. In our applications this proved to be crucial for the acceptance of the presented methods.

Since the proposed approach also includes the feature of including background knowledge into the process, knowledge-rich domains, e.g., the medical domain, directly benefit from this powerful option. For such domains, applying background knowledge is not only sufficient, it is often necessary in order to gain the acceptance of the users: Medical doctors, for example, often expect high-quality results that can only be obtained when enough background knowledge has been provided to the (semi-)automatic discovery and analysis methods.

Furthermore, the power of the method is significantly enhanced when applying background knowledge, since many more causal relations can potentially be inferred by the analysis algorithm. The knowledge can then also be incrementally formalized and captured in the knowledge base for future reference. This feature is especially suitable for less knowledge-rich domains, e.g., specialized technical domains, for which there is only little initial background knowledge available.

A special advantage of the presented method is given by the fact, that the method does not work as a black box: Instead the results and findings of the algorithm, i.e., the identified relations and inferred causal links can always be *explained* by the application of the rules (CCC and CCU) and by the given background knowledge. In this way, a trace of the rule applications can provide significant insight in the case of possible contradictions and erroneous decisions of the algorithm. Then, the user can always refer to these rule applications for locating and fixing potential problems, and also for supporting findings of the algorithm.

6   Conclusion

In this paper, we have presented an approach for knowledge-intensive causal subgroup discovery: We utilize background knowledge for establishing an extended causal model of the domain. The constructed network can then be used to identify potential causal relations, and confounded/effect-modified associations. In a semi-automatic approach, these can then be evaluated and validated by the user. During this step the user is supported by appropriate visualization techniques that are key features for an effective approach. Furthermore, the available background knowledge can be incrementally refined and extended.

In the future, we are planning to integrate an efficient approach for detecting confounding that is directly embedded in the subgroup discovery method. Related work in that direction was described, for example, in [6, 4]. Another interesting direction for future work is given by considering further background knowledge for causal analysis and by integrating existing ontologies into the knowledge base. Additionally, the refinement of the contained relations is an interesting issue to consider.

Acknowledgements

This work has been partially supported by the German Research Council (DFG) under grant Pu 129/8-2.

References

1. Atzmueller, M., Puppe, F.: Semi-Automatic Visual Subgroup Mining using VIKAMINE. Journal of Universal Computer Science 11(11), 1752–1765 (2005)
2. Atzmueller, M., Puppe, F.: SD-Map - A Fast Algorithm for Exhaustive Subgroup Discovery. In: Fürnkranz, J., Scheffer, T., Spiliopoulou, M. (eds.) PKDD 2006. LNCS, vol. 4213, pp. 6–17. Springer, Heidelberg (2006)
3. Atzmueller, M., Puppe, F., Buscher, H.P.: Exploiting Background Knowledge for Knowledge-Intensive Subgroup Discovery. In: Proc. 19th Intl. Joint Conference on Artificial Intelligence (IJCAI 2005), Edinburgh, Scotland, pp. 647–652 (2005)
4. Atzmueller, M., Puppe, F., Buscher, H.P.: A semi-automatic approach for confounding-aware subgroup discovery. International Journal on AI Tools (IJAIT) 18(1), 1–18 (2009)
5. Cooper, G.F.: A Simple Constraint-Based Algorithm for Efficiently Mining Observational Databases for Causal Relationships. Data Min. Knowl. Discov. 1(2), 203–224 (1997)
6. Fabris, C.C., Freitas, A.A.: Discovering Surprising Patterns by Detecting Occurrences of Simpson's Paradox. In: Research and Development in Intelligent Systems, vol. XVI, pp. 148–160. Springer, Berlin (1999)
7. Huettig, M., Buscher, G., Menzel, T., Scheppach, W., Puppe, F., Buscher, H.P.: A Diagnostic Expert System for Structured Reports, Quality Assessment, and Training of Residents in Sonography. Medizinische Klinik 99(3), 117–122 (2004)
8. Kloesgen, W., May, M.: Database Integration of Multirelational Causal Subgroup Mining. Tech. rep. Fraunhofer Institute AIS, Sankt Augustin, Germany (2002)
9. Klösgen, W.: Subgroup Discovery. In: Handbook of Data Mining and Knowledge Discovery, ch. 16.3. Oxford University Press, New York (2002)
10. Lavrac, N., Kavsek, B., Flach, P., Todorovski, L.: Subgroup Discovery with CN2-SD. Journal of Machine Learning Research 5, 153–188 (2004)
11. McNamee, R.: Confounding and Confounders. Occup. Environ. Med. 60, 227–234 (2003)

12. Pearl, J.: Why There is No Statistical Test For Confounding, Why Many Think There Is, and Why They Are Almost Right. In: Causality: Models, Reasoning and Inference, ch. 6.2. Cambridge University Press, Cambridge (2000)
13. Shneiderman, B.: The Eyes Have It: A Task by Data Type Taxonomy for Information Visualizations. In: Proc. IEEE Symp. Visual Languages, pp. 336–343, Boulder, Colorado (1996)
14. Silverstein, C., Brin, S., Motwani, R., Ullman, J.D.: Scalable Techniques for Mining Causal Structures. Data Mining and Knowledge Discovery 4(2/3), 163–192 (2000)
15. Simpson, E.H.: The Interpretation of Interaction in Contingency Tables. Journal of the Royal Statistical Society 18, 238–241 (1951)
16. Webb, G.I.: Discovering Significant Patterns. Machine Learning 68(1), 1–33 (2007)
17. Wrobel, S.: An Algorithm for Multi-Relational Discovery of Subgroups. In: Proc. 1st Europ. Symposion on Principles of Data Mining and Knowledge Discovery, pp. 78–87. Springer, Berlin (1997)

A Study of the SEMINTEC Approach to Frequent Pattern Mining

Joanna Józefowska, Agnieszka Ławrynowicz, and Tomasz Łukaszewski

Abstract. This paper contains the experimental investigation of an approach, named SEMINTEC, to frequent pattern mining in combined knowledge bases represented in description logic with rules (so-called $\mathscr{DL}$-safe ones). Frequent patterns in this approach are the conjunctive queries to a combined knowledge base. In this paper, first, we prove that the approach introduced in our previous work for the DLP fragment of description logic family of languages, is also valid for more expressive languages. Next, we present the experimental results under different settings of the approach, and on knowledge bases of different sizes and complexities.

1 Introduction

Recent developments in knowledge representation and their adoption raise challenges for a research on exploiting new representations in knowledge discovery algorithms. The interest in using explicitly and formally represented *prior (background) knowledge* in knowledge discovery process follows from the fact that it may improve the effectiveness of the process and the quality of its results. However, in most of the current knowledge discovery methods, the background knowledge is implicit or does not have formal structure or semantics and practically can be only considered by the human analyst. An exception are the methods developed within relational data mining (RDM) research. RDM methods belong to the wider group of methods of inductive logic programming (ILP). In ILP, the fragments of first-order logic are used as the language in which data and patterns are represented. RDM approaches usually use Datalog as the representation language. Recently, another logic-based formalism, description logics ($\mathscr{DL}$) [1], has gained an attention. The growing interest in $\mathscr{DL}$ is due to its adoption in the Semantic Web [2] domain,

Joanna Józefowska, Agnieszka Ławrynowicz, and Tomasz Łukaszewski
Institute of Computing Science, Poznan University of Technology,
ul. Piotrowo 2, 60-965 Poznan, Poland
e-mail: {jjozefowska,alawrynowicz,tlukaszewski}@cs.put.poznan.pl

B. Berendt et al. (Eds.): Knowl. Disc. Enhan. with Sem. and Soc. Info., SCI 220, pp. 37–51.
springerlink.com © Springer-Verlag Berlin Heidelberg 2009

currently one of the most active fields of the application of the artificial intelligence research.

A $\mathscr{DL}$ knowledge base typically consists of terminological (schema) part, called *TBox* and assertional (data) part, called *ABox*. Existing $\mathscr{DL}$ reasoners have been mostly developed to efficiently handle complex TBoxes, not focusing on handling large ABoxes. For the data mining task, handling efficiently large datasets is crucial. Only recently the topic of the performance of the query answering over knowledge bases with large ABoxes has gained more attention. Recently implemented reasoning engine KAON2[1] outperforms other reasoners in case of a high number of instances in the knowledge base and not very complex TBox [18] (for some other tests of the reasoners see also [22]). KAON2 uses an algorithm that transforms a knowledge base *KB* in $\mathscr{DL}$ to a knowledge base in Disjunctive Datalog DD(*KB*) [8]. This transformation enables using optimization techniques that proved to be effective in disjunctive databases, such as magic-sets or join-order optimizations.

With regard to expressiveness, $\mathscr{DL}$ offers complementary features to Datalog (or Logic Programming in general) [4]. Datalog allows relations of arbitrary arity and arbitrary composition of relations, but it does not allow to use the existential quantifier. $\mathscr{DL}$ allows both quantifiers, existential and universal one, and provides full, monotonic negation. Unfortunately the axioms it can express are only of tree-like structure. Additionally, it is restricted to only unary and binary predicates. The combination of the expressive power of $\mathscr{DL}$ and Datalog would be thus desirable. Abovementioned KAON2 reasoner supports so-called $\mathscr{DL}$-*safe rules* [17], a combination of $\mathscr{DL}$ and function-free Horn rules, where very expressive $\mathscr{DL}$ can be used, while preserving the *decidability* property.

In [9] we discussed the potential of using this combination, and the query answering method implemented in KAON2, for frequent pattern discovery in knowledge bases represented in $\mathscr{DL}$ with rules. In [10] we presented an approach to frequent pattern mining in knowledge bases in DLP [7], the language restricted to the intersection of the expressivity of $\mathscr{DL}$ and Logic Programming. In this paper we prove that our method for building patterns remains correct for more expressive languages from $\mathscr{DL}$ family. We discuss different settings that we implemented under our approach. The results of the experimental investigation under these settings, for different kinds of knowledge bases, are presented. We will refer to our proposed approach as to SEMINTEC.

The rest of the paper is organized as follows. In Section 2 we present the related work. In Section 3 we present the data mining setting. Section 4 contains the overview of our approach. In Section 5 the experimental results are presented, and Section 6 concludes the paper.

2   Related Work

The ILP methods, in particular RDM ones, are closely related to our work. WARMR [5] is the first relational data mining system proposed for frequent pattern discovery.

---

[1] http://kaon2.semanticweb.org

The initial approach was further refined and optimized [3]. Another RDM system, FARMER [19, 20], instead of ILP techniques uses only first-order logic notation and efficient *trie* data structure. Encouraged by the results achieved by FARMER we decided to use similar structure in our approach. The aforementioned methods use, as a generality relation, θ-subsumption which is a purely syntactic measure. In turn, system c-armr [21], also an ILP frequent query miner, implements the refinement operator under generalized subsumption, which is a semantic measure. C-armr induces frequent queries which do not contain redundant literals, where redundancy is defined relative to a background theory. Our approach is also based on the generalized subsumption and exploits the background knowledge for pruning redundant patterns.

To the best of our knowledge, there has been only one approach, named SPADA [13, 14], that uses a hybrid language combining description logic with rules as the representation language for frequent pattern mining. More specifically, SPADA, and its further versions under the name $\mathcal{AL}$-QuIn [12, 11], use $\mathcal{AL}$-log [6] language, which combines $\mathcal{ALC}$ language from $\mathcal{DL}$ family with Datalog rules. In these rules, there can be $\mathcal{ALC}$ concepts as predicates in the unary atom constraints in the body. Any roles from $\mathcal{DL}$ knowledge base may not occur there. However, in contrast to the approach presented in this work, patterns in SPADA/$\mathcal{AL}$-QuIn can contain n-ary Datalog predicates. The $\mathcal{DL}$-safe rules combination supports more expressive $\mathcal{DL}$ than $\mathcal{AL}$-log, and allows using both, concepts and roles in atoms. Concepts and roles may also be used in rule heads. Query answering algorithm, proposed as the main reasoning technique for $\mathcal{DL}$-safe rules, is based on deductive database techniques [17, 18]. It runs in EXP time, while the algorithm for $\mathcal{AL}$-log runs in single NEXP time. The algorithm is based on the reduction of $\mathcal{DL}$ knowledge base into a knowledge base represented in positive Disjunctive Datalog [8]. Thus, the internal representation in our approach is positive Disjunctive Datalog, while in SPADA/$\mathcal{AL}$-QuIn it is Datalog. In SPADA/$\mathcal{AL}$-QuIn, the notion of description granularity is exploited, what means that the patterns contain the concepts from given levels of a concept taxonomy. In our approach, a pattern may contain concepts from different levels of a taxonomy.

3 Problem Statement

3.1 Pattern Discovery Task

The general formulation of the frequent pattern discovery problem was specified by [15]. It was further extended to deal with more expressive language in case of RDM methods in [5]. With respect to these formulations we define our task as:

Definition 1. Given

- a combined knowledge base (KB, P), containing $\mathcal{DL}$ component KB and a logic program containing rules P,

- a set of patterns in the form of queries Q that all contain a reference concept $\hat{C}$ as a predicate in one of the atoms in the body and where the variable in the atom with $\hat{C}$ is the only distinguished one in the query Q,
- a minimum support threshold *minsup* specified by the user

and assuming that queries with support s are frequent in (KB,P) given $\hat{C}$ if $s \geq$ *minsup*, the task of frequent pattern discovery is to find the set $\mathscr{F}$ of frequent queries.

The parameter $\hat{C}$, a reference concept, determines what is counted during the calculation of the support of a given query. An atom of the query with reference concept $\hat{C}$ as a predicate contains the only one distinguished variable in the query, called *key*. The support is calculated as the ratio between the number of the bindings of variable in the given query and the number of the bindings of variable *key* in the reference query.

Definition 2. A support of the query Q with respect to the knowledge base (KB,P) is defined as the ratio between the number of instances of the $\hat{C}$ concept that satisfy the query Q and the total number of instances of the $\hat{C}$ concept (the trivial query for the total number of the instances is denoted Q_{ref}):

$$support(\hat{C},Q,(KB,P)) = \frac{|answerset(\hat{C},Q,(KB,P))|}{|answerset(\hat{C},Q_{ref},(KB,P))|} \tag{1}$$

3.2 The Data Mining Setting

Our method assumes mining patterns in a combined knowledge base (KB,P), where KB is description logic component and P is a program containing a set of (disjunctive) rules, so-called $\mathscr{DL}$-safe ones [17].

The basic building blocks of $\mathscr{DL}$ KB are *concepts*, (representing sets of objects), *roles*, (representing relationships between objects), and *individuals*, representing objects itselves. Complex concepts are defined using the atomic ones, roles and a set of constructors such as $\sqcap$ (conjunction), $\sqcup$ (disjunction), and $\neg$ (negation). A $\mathscr{DL}$ KB typically consists of a *TBox* with assertions about concepts such as subsumption (*Woman* $\sqsubseteq$ *Person*) and *ABox* with role assertions between individuals (*livesIn*(*Anna*,*Europe*)) and membership assertions (*Woman*(*Anna*)). The semantics of a $\mathscr{DL}$ KB is given by translating it into first order logics. Atomic concepts and roles are translated into unary and binary predicates, and complex concepts into formulae with one free variable.

In a $\mathscr{DL}$-safe rule each variable occurs in some non-DL-atom in the rule body (that is an atom from the ones outside KB). It makes the rule applicable only to individuals explicitly introduced to a knowledge base.

The semantics of the combined knowledge base (KB,P) in the $\mathscr{DL}$-safe rules approach is given by the translation into the first-order logic as $\pi(KB) \cup P$. For the details of the transformation π we refer the reader to [17].

The formalism used in our approach is, in general, that of $\mathscr{DL}$-safe rules, introduced in [17]. However, there are some restrictions imposed. In rules as well as in queries, we assume only $\mathscr{DL}$-atoms as literals. That is, only concepts and roles can be used as predicates, but any n-ry predicates from P. The subset of description logic considered is restricted to $\mathscr{SHIF}$. In the work introducing $\mathscr{DL}$-safe rules [17], there is the subset $\mathscr{SHOIN}(\mathbf{D})$ assumed. $\mathscr{SHOIN}(\mathbf{D})$ corresponds to OWL DL variant of OWL language[2], a standard ontology language for the Web, while $\mathscr{SHIF}(\mathbf{D})$ to its lighter version called OWL-Lite. $\mathscr{SHIF}$ is $\mathscr{SHIF}(\mathbf{D})$ without datatypes, which is another restriction of the language assumed by the method presented in this work. Hovever, datatypes are planned to be integrated in the future work.

In our approach the intensional background knowledge is represented in a TBox. An ABox contains instances (extensional background knowledge). Our goal is to find frequent patterns in the form of conjunctive, positive $\mathscr{DL}$-safe queries over (KB, P). $\mathscr{DL}$-safe queries are conjunctive queries without true non-distinguished variables. This makes them applicable only to the explicitly introduced individuals. All variables in such a query are bound to individuals explicitly occurring in a knowledge base, even if they are not returned as part of the query answer. More specifically, the frequent pattern $\mathscr{Q}$ has the form of conjunctive, positive $\mathscr{DL}$-safe query over (KB, P), whose answer set contains individuals of concept $\hat{C}$.

Definition 3 (Pattern). Given is a combined knowledge base (KB, P). A pattern Q is a conjunctive, positive $\mathscr{DL}$-safe query over (KB, P) of the following form:

$$Q(key) =? - \hat{C}(key), B_1, ..., B_n, \mathscr{O}(key), \mathscr{O}(x_1), ..., \mathscr{O}(x_m)$$

where $B_1, ..., B_n$ represent atoms of the query (where predicates are either atomic concepts or simple roles), $Q(key)$ denotes that variable key is the only one variable whose bindings are returned in the answer (distinguished variable), $x_1, ..., x_m$ represent the variables of the query which are not a part of the answer (existential variables).

A trivial pattern is a query of the form:

$$Q(key) =? - \hat{C}(key), \mathscr{O}(key)$$

Special predicates $\mathscr{O}(v)$ indicate that queries are $\mathscr{DL}$-safe, that is all variables are bound to individuals explicitly present in the knowledge base. Further in the paper we omit the special predicates and assume that all the considered queries are $\mathscr{DL}$-safe. For *Client* being reference concept $\hat{C}$ the following example query can be build:

$$Q(key) =? - Client(key), isOwnerOf(key, x), Account(x), hasLoan(x, y), Loan(y)$$

We assume to use a semantic generality relation $\succeq_{\mathscr{B}}$ to assure that the chosen relation does not disregard the presence of a combined knowledge base (KB, P). In our method, patterns are represented as queries. Therefore, we propose to define the generality relation (or measure) between two patterns as *query containment* (or *subsumption*) relation.

---

[2] http://www.w3.org/TR/owl-features/

Definition 4 (Query containment). Given two queries Q_1 and Q_2 to a knowledge base (KB,P) we say that Q_1 is at least as general as Q_2 under query containment, $Q_1 \succeq_{\mathscr{B}} Q_2$, iff in every possible state of the (KB,P) the answer set of the Q_2 is contained in the answer set of the Q_1.

Due to the definition of the query support it is easy to prove that the query containment is monotonic w.r.t. the support.

3.3 The Correctness of the Pattern Refinement Method for $\mathscr{SHIF}$

In general, more specific query Q_2 is built from more general one Q_1 by adding an atom to the query Q_1. The correctness of such pattern refinement method for a combined knowledge base (KB,P) is discussed next. Recall, that the method for query answering used in our approach is based on the translation of a description logic knowledge base KB into a Disjunctive Datalog DD(KB). The difference between the general first-order semantics of $\mathscr{DL}$ and minimal model semantics of Disjuctive Datalog is not relevant for answering positive queries in positive Datalog programs, because all positive ground facts entailed by KB are contained in each minimal model of DD(KB) and vice versa. The reduction to DD(KB) produces only positive programs [8]. The rules in program P are simply appended to the result of such reduction. A positive Disjunctive Datalog program DD(KB)$\cup P$ is obtained.

Lemma 1. *Let Q_2 be a query over KB, built from the query Q_1 by adding an atom. It holds that $Q_1 \succeq_{\mathscr{B}} Q_2$.*

Proof (Sketch). If the program in Disjunctive Datalog remains fixed, also the answer sets stay fixed. The program DD(KB) is independent of the query, as long as the query is built from positive atomic concepts or roles (see: [17]). Queries in our approach, do not contain any atoms with complex concepts or roles (only with atomic and simple ones), thus the query does not introduce any new symbols to a TBox and the reduction is independent of the query. Hence, DD(KB) can be computed once and remains fixed while answering to any number of the queries of the described form. Thus adding positive literals to the query can at most reduce answers and the thesis follows.

3.4 Illustrative Example

As an illustrative example within this paper let us consider the knowledge base describing bank services and its clients.

Example 1. The TBox of KB and rules are presented below. The knowledge base contains axioms like domain and range specifications of some roles (*isOwnerOf* and *hasLoan*), disjointness constraints between some concepts (e.g. *Man* and *Woman*) and the concept definition (*Interesting*).

$\top \sqsubseteq \forall isOwnerOf.(Account \sqcup CreditCard)$ $\top \sqsubseteq \forall hasLoan.Loan$
$\top \sqsubseteq \forall isOwnerOf^-.Client$ $\top \sqsubseteq \forall hasLoan^-.Account$
$\top \sqsubseteq\ \leq 1\ isOwnerOf^-$ $Interesting \equiv \exists hasLoan.Loan$

$Account \equiv \neg Client$ $Client(x) \leftarrow Man(x)$
$Account \equiv \neg CreditCard$ $Client(x) \leftarrow Woman(x)$
$Account \equiv \neg Loan$ $CreditCard(x) \leftarrow Gold(x)$
$Client \equiv \neg CreditCard$
$Client \equiv \neg Loan$
$CreditCard \equiv \neg Loan$
$Man \equiv \neg Woman$

In the ABox there are the following assertions:

Man(Marek).	*Account(a1).*	*isOwnerOf(Marek, a1).*
Man(Adam).	*Account(a2).*	*isOwnerOf(Marek, c1).*
Woman(Anna).	*Account(a3).*	*isOwnerOf(Anna, a2).*
Woman(Maria).	*CreditCard(c1).*	*isOwnerOf(Anna, c2).*
	CreditCard(c2).	*isOwnerOf(Maria, a3).*

Let us assume that the reference concept is *Client*. Then the query Q_{ref} has the form

$$Q(key) =? - Client(key)$$

and has 4 items in its answerset. Let us assume further that we would like to calculate the support of the example query Q of the form:

$$Q(key) \quad =? \quad - \quad Client(key), isOwnerOf(key, x), Account(x), isOwnerOf(key, y), CreditCard(y)$$

The query Q has two items in its answerset that are the clients having at least one account and at least one credit card. The support of the query Q is then calculated as: $support(\hat{C}, Q, KB) = \frac{2}{4} = 0.5$.

4 Overview of the Approach

The desirable feature of a generality relation, crucial for developing efficient algorithms is its monotonicity with regard to support. This feature is present in our case. More specifically, our algorithms are based on the property of the query support that for every pair of patterns $p1$ and $p2$: $p1 \succeq p2 \Rightarrow support(p1) \geq support(p2)$. It can be thus apriori determined that more specific patterns subsumed by an infrequent pattern are also infrequent.

We use the special *trie* data structure that was successfully used in FARMER. In the trie data structure, nodes correspond to the atoms of the query. Every path from the root to a node corresponds to a query (see Figure 2). New nodes are added to the trie, only if the resulting queries are frequent. Thus only leaves that correspond to frequent queries are expanded. We distinguish two ways in which atoms can be added as leaves to the trie, as described in Definition 5.

Definition 5. Atoms are added to the trie as:

1. **dependent atoms** (use at least one variable of the last atom in the query),
2. **right brothers** of a given node (these are the copies of atoms that have the same parent node that a given node and are placed on the right side of a given node).

Depenedent atoms are built from predicates from the *admissible predicates* list. This list is computed for each predicate, when it is added in some atom to the trie for the first time. It contains also the information which variables of the child node are to be shared with the parent node (a predicate may be added in several ways as admissible one, also to itself). The list is then stored in a hash structure and retrieved when the atom with the given predicate is expanded for the next time. To add a predicate to the list, the intersections of descriptions (from parent and child predicate) describing future shared variables have to be satisfiable w.r.t. the TBox.

Right brother copying mechanism takes care that all possible subsets, but only one permutation out of a set of dependent atoms is considered. Variables in binary predicates are renamed, where needed, while being copied.

4.1   The Implemented Settings

We implemented two versions of the approach:

- generating all *semantically closed patterns* (that is the patterns to which it is not possible to add any literal without effecting the semantics. These are the largest patterns in a class of semantically equivalent patterns) and,
- generating all *semantic equivalence classes of patterns* (that is at least one representative of each class of semantically equivalent patterns, but here we are interested in generating possibly the shortest patterns).

An example closed pattern in Figure 2 is

$$Q(key) =? - Client(key), isOwnerOf(key,x0), CreditCard(x0)$$

An example representative of an equivalence class is

$$Q(key) =? - Client(key), isOwnerOf(key,x0), Gold(x0)$$

(to make it closed the literal $CreditCard(x0)$ should be added). For each one of these two versions, we apply some techniques, based on syntax and semantics of patterns to generate all patterns and eliminate redundant patterns of each kind.

Some techniques are based only on the syntactic form of a pattern, without using the semantic information from a TBox. These operations rely on introducing new variable names systematically, level by level, what helps to maintain the hash lists of previously added dependent atom forms. As the atoms of the form already generated will be copied as right brothers, they should not be generated again in the same form (different only due to the variable renaming) as dependent atoms in the next level.

On the semantic part, the method starts from classifying the concept taxonomy, which, together with properties hierarchy, serves for regular construction of a trie. In case of closed patterns, to admissible predicates list the concepts and properties from the top level of hierarchies are added, and their direct subconcepts/subproperties are added to them on the next level (see: Figure 2a, where only the top concepts *CreditCard* and *Account* were considered as admissible predicates of *isOwnerOf*). In case of equivalence classes, the whole branches from the top to the bottom of a hierarchy are added at the same level (see: Figure 2b). The subsumption hierarchy is used in several ways. Consider, for example, the situation where the domain/range of a given property is equivalent to some class to be added as an admissible predicate to this property. In case of equivalence classes, we do not add this class as it does not bring any new semantic information (for example: $Loan(x1)$ would not be added to $hasLoan(x0, x1)$). The attributes of properties (symmetric, inverse) are treated differently in the two settings, e.g. for closed patterns we generate both, property and its inverse, while for equivalence classes, the information that one property is an inverse of another one helps to prune semantically redundant literals where possible. For transitive properties, the transitive closure is generated while constructing closed patterns. In this work we do not check whether a new pattern is not semantically equivalent to one of the previously generated ones (which would be expensive), what results in generating redundant patterns. However, this is planned for the future versions of our approach.

Before submitting a query to calculate its support, we test if the constructed query is satisfiable w.r.t. a TBox T, to eliminate unnecessary computation w.r.t. the data base. We decided to test two methods: *complete test of query satisfiability* and its *approximation*. First test consists in checking whether $T \cup \exists \mathbf{x}, \mathbf{y} : Q$ is satisfiable, that is, whether there is a model of T in which there is some valuation for the distinguished variables $\mathbf{x}$ and nondistinguished variables $\mathbf{y}$. The variables are skolemized, and, assuming that $Q(\mathbf{a}, \mathbf{b})$ is a new ABox, it is checked whether T is satisfiable in conjunction with that ABox. In the second test, for each variable in a query, a description is built as an intersection of all descriptions from concepts, domains and ranges of properties describing this variable in a query. The descriptions are kept on the hash list associated with every node and updated for new atom being added to the query (see: Figure 2a). It is checked whether the intersection of the descriptions of the shared variables from a new atom and from a given query, to which we are going to add this atom, are satisfiable.

Below we present the general node expansion algorithm[3]. $P(\mathbf{x}, \mathbf{y})$ denotes an atom where P is a predicate name and $\mathbf{x}$ and $\mathbf{y}$ distinguished and undistinguished variables. The trie is generated up to the user-specified MAXDEPTH level.

---

[3] In case of closed patterns there is an additional step of generating transitive closure for transitive properties, which is not present in this algorithm.

Algorithm 1. *expandNode(A(x, y_a), nodeLevel)*

1. if nodeLevel < MAXDEPTH then
2.   if admissible predicates of A not computed then
3.   computeAdmissiblePredicates(A);
4.   for all D ∈ admissible predicates of A do
5.   build dependent atom $D(x, y_d)$ of $A(x, y_a)$
6.   if candidate query Q satisfiable then
7.   if candidate query Q frequent then
8.   addChild($A(x, y_a)$, $D(x, y_d)$);
9.   for all $B(x, y_b)$ ∈ right brothers of $A(x, y_a)$ do
10.   create $B'(x, y_{b'})$ which is a copy of node $B(x, y_b)$;
11.   if candidate query Q satisfiable then
12.   if candidate query Q frequent then
13.   addChild($A(x, y_a)$, $B'(x, y_{b'})$);
14.   for all $C(x, y_c)$ ∈ children of $A(x, y_a)$ do
15.   expandNode($C(x, y_c)$, nodeLevel+1)

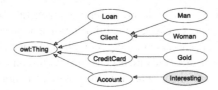

Fig. 1 Classified taxonomy

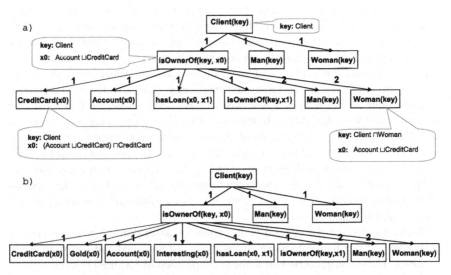

Fig. 2 A part of a trie generated for *Client* as a reference concept. (a) sematically closed patterns. (b) semantic equivalence classes of patterns.

Example 2. (Following Example 1). Let us assume having the TBox from Example 1 and the ABox bigger than in previous example (for the sake of clarity we will not discuss it within this example). Then our method works as follows. First we classify a taxonomy and as an effect we obtain the classification presented in Figure 1. The top-level concepts in the example are: *CreditCard*, *Client*, *Account*, *Loan*. Concept *Interesting* is the only one concept that changed place in the taxonomy after the classification. Originally it was on top of the taxonomy, after the classification it became a child of *Account*. For the predicate *Client* admissible predicates are: *isOwnerOf* (only first variable can be the shared one), *Man* and *Woman*. An example trie is presented in Figure 2. The numbers on edges refer to two ways in which the atoms can be added to the trie. Some of the hash lists of variable descriptions associated with nodes are shown in Figure 2a. The trie is built up to the level 3 (to the patterns having the length of 3 atoms). The example pattern at this level is: $Q(key) =? - Client(key), isOwnerOf(key, x0), CreditCard(x0)$.

5 Experimental Results

We implemented and tested all of the settings of our SEMINTEC approach, discussed in the previous section. The primary goal of the experiments was to estimate the increase of the data mining efficiency by using background knowledge from a TBox in different settings. We wanted also to test how our method performs on datasets of different sizes and complexities (within the OWL Lite fragment), to obtain an idea what kinds of ontologies can be efficiently handled.

To test our approach, we created the ontology based on the financial dataset from the PKDD'99 Discovery Challenge (*PKDD CUP*)[4]. It is relatively simple, as it does not use existential quantifiers or disjunctions. It requires, however, hard for deductive databases, equality reasoning, as it contains functionality assertions and disjointness constraints.

LUBM is a benchmark from the Lehigh University[5], consisting of a university domain ontology and a generator of a synthetic data (we set the number of universities to 1). There are existential quantifiers used, but no disjunctions or number restrictions, hence the reduction algorithm produces an equality-free Horn program, on which query answering can be performed deterministically.

The *Biopax* ontology contains biological pathway data[6]. For tests we used Agro-Cyc dataset. It uses existential quantifiers, disjunctions, functionality assertions and disjointness constraints. Since it contains also nominals, which KAON2 cannot handle, we adapted it for our tests. Namely, each enumerated concept $i_1, ..., i_n$ was replaced with a new concept O and its subconcepts $O_1, ..., O_n$ and assertions $O_k(i_k)$ added. The datatype properties P_d, used in axioms, were replaced by object properties P_o and the statements of the form $\exists P_d.\{i_k\}$ with the form $\exists P_o.O_k$.

[4] PKDD CUP, http://www.cs.put.poznan.pl/alawrynowicz/goldDLP2.owl
[5] LUBM, http://swat.cse.lehigh.edu/projects/lubm/
[6] BioCyc Database Collection, http://biocyc.org

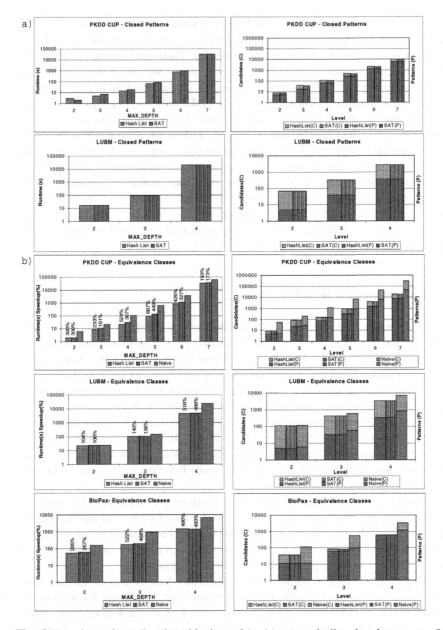

Fig. 3 Experimental results (logarithmic scale). (a) semantically closed patterns. (b) semantic equivalence classes. **HashList**: an approximation of query satisfiability test. **SAT**: complete test. **Naive**: naive approach.

In Figure 3 some of our experimental results are presented. The tests were performed on a computer with Intel Core2 Duo 2.4GHz processor, 2GB of RAM, running Microsoft Windows Serwer 2003 Standard Edition SP1. Our implementation and KAON2 (release 2006-12-04) are in Java (version 1.5.0). The JVM heap size was limited to 1.5GB, the maximum time for each test to 25 hours. A trie was generated up to specified MAXDEPTH values, from 1 to the maximum depth where the time of 25 hours was not exceeded, and using the recursive strategy presented in the Algorithm 1. The runtimes are the times of a whole trie generation for each MAXDEPTH (maximum length of patterns). However, the numbers of candidates and frequent patterns found are shown separately for each level (each length of patterns). Besides the runtimes, there is also the speedup presented of the better methods compared to the naive approach. The bars representing the numbers of patterns are superimposed on top of the ones representing the numbers of candidates. For PKDD CUP we set the *minsup* to *0.2* and the $\hat{C}$ concept to *CreditCard*. For LUBM and BioPax we set the *minsup* to *0.3* and the $\hat{C}$ concepts adequatly to *Person* and to *entity*.

An example pattern discovered, from the PKDD CUP dataset, is: *Q(key)= ?- Client(key), isOwnerOf(key,x1), NoProblemAccount(x1), hasStatementIssuanceFrequency(x1,x2), Monthly(x2), hasAgeValue(key,x3), From35To50(x3)* (support: 0,23). It describes the clients of the age between 35 and 50 years, having at least one account with the statements issued monthly and no problem status of paying off the loans (i.e. any loan granted or only no problem loans). The latter information can be read from the PKDD CUP ontology, where *NoProblemAccount* is defined as *Account* having only no problem loans (*NoProblemAccount ⊑ Account ⊓ ∀hasLoan.OKLoan*). Using background knowledge from the TBox, in the equivalence classes case, saved us from generating the patterns where both, *AgeValue* and *From35To50* or *hasAgeValue* and *AgeValue* predicates, are present (the following axioms were used: *From35To50 ⊑ AgeValue* and ⊤ ⊑ ∀ *hasAgeValue.AgeValue*).

In the naive approach, where no semantics is exploited, we applied all the implemented techniques based on the syntax of patterns for elimination of redundant patterns. Nevertheless, the naive approach still generated many more patterns, to be tested, at each level. *HashList* approximation for testing query satisfiability was in most cases faster than the complete test (*SAT*), even though it had to evaluate more queries. Testing on LUBM showed an important feature of our approach. There are no disjointness constraints in this ontology, and hence we cannot eliminate too many from the admissible predicates lists, what causes a lot of atoms to be tested each time as dependent ones. However, when using semantic information (such as the subsumption hierarchy of concepts), there are speedups even for this ontology. For the other two ontologies, the presence of disjointness constraints helped to eliminate many useless literals before their actual evaluation. In LUBM there are also the transitive roles, what causes big shift in the runtime in the closed patterns case.

6 Conclusion and Future Work

In this paper we present an evaluation of the approach, named SEMINTEC, to frequent pattern mining in combined knowledge bases consisting of $\mathcal{DL}$ component and $\mathcal{DL}$-safe rules. After introducing the general algorithm, based on the trie data structure, we discuss different settings under this algorithm. We present the experimental results of testing the presented settings on knowledge bases of different sizes and complexities.

To the best of our knowledge, there is only one other approach for frequent pattern discovery, system SPADA/$\mathcal{AL}$-QuIn, that uses $\mathcal{DL}$ to represent background knowledge. We compared qualitatively some of the features of SPADA/$\mathcal{AL}$-QuIn and SEMINTEC. However, it may be difficult to perform direct quantitative comparison due to different languages and different forms of patterns considered.

Further research includes development within our approach of methods for pruning patterns semantically equivalent to already found ones, by generating only semantically free patterns and search through the trie. Furthermore, in the future we assume to integrate n-ary predicates to be allowed in both: rules and queries (patterns).

In the future we plan to focus on optimization techniques and heuristic algorithms to speed up pattern discovery process. We also plan to investigate the measures of interestingness with the goal of pruning huge pattern space to leave only interesting patterns.

References

1. Baader, F., Calvanese, D., McGuinness, D., Nardi, D., Patel-Schneider, P. (eds.): The description logic handbook: Theory, implementation and applications. Cambridge University Press, Cambridge (2003)
2. Berners-Lee, T., Hendler, J., Lassila, O.: The Semantic Web. Scientific American 284(5), 34–43 (2001)
3. Blockeel, H., Dehaspe, L., Demoen, B., Janssens, G., Ramon, J., Vandecasteele, H.: Improving the efficiency of Inductive Logic Programming through the use of query packs. Journal of Artificial Intelligence Research 16, 135–166 (2002)
4. Borgida, A.: On the relative expressiveness of description logics and predicate logics. Artificial Intelligence 82(1/2), 353–367 (1996)
5. Dehaspe, L., Toivonen, H.: Discovery of frequent Datalog patterns. Data Mining and Knowledge Discovery 3(1), 7–36 (1999)
6. Donini, F., Lenzerini, M., Nardi, D., Schaerf, A.: AL-log: Integrating datalog and description logics. Journal of Intelligent Information Systems 10(3), 227–252 (1998)
7. Grosof, B.N., Horrocks, I., Volz, R., Decker, S.: Description Logic Programs: Combining Logic Programs with Description Logic. In: Proc. of the Twelfth Int'l World Wide Web Conf (WWW 2003), pp. 48–57. ACM, New York (2003)
8. Hustadt, U., Motik, B., Sattler, U.: Reducing $\mathcal{SHIQ}$ Description Logic to Disjunctive Datalog Programs. In: Proc. of KR 2004, pp. 152–162. AAAI Press, Menlo Park (2004)
9. Józefowska, J., Ławrynowicz, A., Łukaszewski, T.: Towards discovery of frequent patterns in description logics with rules. In: Adi, A., Stoutenburg, S., Tabet, S. (eds.) RuleML 2005, vol. 3791, pp. 84–97. Springer, Heidelberg (2005)
10. Józefowska, J., Ławrynowicz, A., Łukaszewski, T.: Frequent pattern discovery in OWL DLP knowledge bases. In: Staab, S., Svátek, V. (eds.) EKAW 2006. LNCS (LNAI), vol. 4248, pp. 287–302. Springer, Heidelberg (2006)
11. Lisi, F.A., Esposito, F.: Efficient Evaluation of Candidate Hypotheses in $\mathcal{AL}$-log. In: Camacho, R., King, R., Srinivasan, A. (eds.) ILP 2004. LNCS (LNAI), vol. 3194, pp. 216–233. Springer, Heidelberg (2004)

12. Lisi, F.A., Malerba, D.: Ideal Refinement of Descriptions in $\mathcal{AL}$-log. In: Horváth, T., Yamamoto, A. (eds.) ILP 2003. LNCS (LNAI), vol. 2835, pp. 215–232. Springer, Heidelberg (2003)
13. Lisi, F.A., Malerba, D.: Inducing Multi-Level Association Rules from Multiple Relation. Machine Learning Journal 55, 175–210 (2004)
14. Lisi, F.A., Esposito, F.: An ILP Approach to Semantic Web Mining. In: Proc. of Knowledge Discovery and Ontologies (KDO-2004), Workshop at ECML/PKDD 2004 (2004)
15. Mannila, H., Toivonen, H.: Levelwise search and borders of theories in knowledge discovery. Data Mining and Knowledge Discovery 1(3), 241–258 (1997)
16. Motik, B., Rosati, R.: A Faithful Integration of Description Logics with Logic Programming. In: Proc. of IJCAI 2007, pp. 477–482. Morgan Kaufmann, San Francisco (2007)
17. Motik, B., Sattler, U., Studer, R.: Query Answering for OWL-DL with Rules. In: McIlraith, S.A., Plexousakis, D., van Harmelen, F. (eds.) ISWC 2004. LNCS, vol. 3298, pp. 549–563. Springer, Heidelberg (2004)
18. Motik, B., Sattler, U.: A Comparison of Reasoning Techniques for Querying Large Description Logic ABoxes. In: Hermann, M., Voronkov, A. (eds.) LPAR 2006. LNCS, vol. 4246, pp. 227–241. Springer, Heidelberg (2006)
19. Nijssen, S., Kok, J.N.: Faster Association Rules for Multiple Relations. In: Proc. of IJCAI 2001, pp. 891–897. Morgan Kaufmann, San Francisco (2001)
20. Nijssen, S., Kok, J.N.: Efficient frequent query discovery in FARMER. In: Lavrač, N., Gamberger, D., Todorovski, L., Blockeel, H. (eds.) PKDD 2003. LNCS (LNAI), vol. 2838, pp. 350–362. Springer, Heidelberg (2003)
21. De Raedt, L., Ramon, J.: Condensed representations for Inductive Logic Programming. In: Principles of Knowledge Representation and Reasoning, Proceedings of the KR 2004, pp. 438–446 (2004)
22. Weithoener, T., Liebig, T., Luther, M., Boehm, S., von Henke, F., Noppens, O.: Real-World Reasoning with OWL. In: Franconi, E., Kifer, M., May, W. (eds.) ESWC 2007. LNCS, vol. 4519, pp. 296–310. Springer, Heidelberg (2007)

Partitional Conceptual Clustering of Web Resources Annotated with Ontology Languages

Floriana Esposito, Nicola Fanizzi, and Claudia d'Amato

Abstract. The paper deals with the problem of cluster discovery in the context of Semantic Web knowledge bases. A partitional clustering algorithm is presented. It is applied for grouping resources contained in knowledge bases and expressed in the standard ontology languages. The method exploits a language-independent semi-distance measure for individuals that is based on the semantics of the resources w.r.t. a context represented by a set of concept descriptions (discriminating features). The clustering algorithm adapts BISECTING K-MEANS method to work with medoids. Besides, we propose simple mechanisms to assign each cluster an intensional definition that may suggest new concepts for the knowledge base (*vivification*). A final experiment demonstrates the validity of the approach through absolute quality indices for clustering results.

1 Unsupervised Learning Based on Similarity Measures

In the inherently distributed applications related to the *Semantic Web* [2] (henceforth SW) there is an extreme need of automatizing those activities which turn out to be more complex and burdensome for the knowledge engineer, such as ontology construction, matching and evolution. Such an automation may be assisted by crafting supervised or unsupervised methods for the specific SW representations (RDF through OWL [9]) founded in *Description Logics* (DLs) [1].

In this work, we investigate on unsupervised learning for knowledge bases expressed in such standard concept languages. In particular, we focus on

Floriana Esposito, Nicola Fanizzi, and Claudia d'Amato
Dipartimento di Informatica – Università degli Studi di Bari
Campus Universitario, Via Orabona 4, 70125 Bari, Italy
e-mail: {esposito,fanizzi,claudia.damato}@di.uniba.it

B. Berendt et al. (Eds.): Knowl. Disc. Enhan. with Sem. and Soc. Info., SCI 220, pp. 53–70.
springerlink.com   © Springer-Verlag Berlin Heidelberg 2009

the problem of *conceptual clustering* [28] of semantically annotated resources
that consists in discovering and intentionally defining meaningful groups
of unlabeled (namely unclassified) resources. This is because, the result
of a conceptual clustering process can be exploited for improving the effi-
ciency/effectiveness of several tasks. For instance:

- clustering annotated resources enables the definition of new emerging concepts (*concept formation*) on the grounds of the primitive concepts as-
  serted in a knowledge base [12];
- supervised methods can exploit the obtained clusters to induce new concept definitions or to refining existing ones (*ontology evolution*);
- intensionally defined groups may be used for speeding-up the search and
  discovery of resources [8];
- a hierarchy, result of a hierarchical clustering process, can be exploited as
  criterion for *ranking* retrieved resources.

The goal of clustering methods is to organize collections of unlabeled resources into meaningful clusters such that their intra-cluster similarity (similarity among resources in the same cluster) is high and their inter-cluster
similarity (similarity among resources belonging to different clusters) is low.
To accomplish this goal, a notion of similarity (or density) is generally exploited. However, these methods rarely take into account forms of *background
knowledge* in order to characterize cluster configurations, for example for giv-
ing a definition of the obtained clusters in terms of the notions (concepts
and relationships) defined in a background knowledge. They simply discover
homogeneous groups of resources. This hinders the interpretation of the clusters, that is crucial in the SW perspective which foresees sharing and reusing
knowledge to enable semantic interoperability. For this reason, conceptual
clustering methods have been developed; their goal is to supply intentional
description of the discovered clusters. This is a difficult problem and few
works have focused such a problem. One of the most important conceptual
clustering algorithm has been proposed by Stepp and Michalski in [28]. Here,
intentional cluster descriptions are determined as conjunctive descriptions
of attributes selected among those that describe the clustered objects [28].
Related works to conceptual clustering methods have been dedicated to similarity measures for clausal spaces [25], instance based classification [10] and
clustering [20]. Yet these representation are known to be incomparable with
DLs [5]. Hence, the application of clustering techniques to semantically annotated resources requires new methods for defining intentional cluster descriptions and new similarity measures able to exploiting the expressiveness
of DL language.

As pointed out in a seminal paper on similarity measures for DLs [6], most
of the existing measures focus on the similarity of atomic concepts within hierarchies or simple ontologies. Moreover, they have been conceived for assessing
concept similarity, whereas, for other tasks, a notion of similarity between *individuals* is required. Recently, dissimilarity measures for specific DLs have

been proposed [7, 17]. Although they resulted to be quite effective [7], they are still partly based on structural criteria which makes them fail to fully grasp the underlying semantics and hardly scale to the complexity of the standard ontology languages. Therefore, we have devised a family of dissimilarity measures for semantically annotated resources, which can overcome the aforementioned limitations.

Following the criterion of semantic discernibility of individuals, we present a new family of measures that is suitable for a wide range of languages since it is merely based on the discernibility of the input individuals with respect to a fixed set of features (henceforth a *committee*) represented by concept definitions. As such the new measures are not absolute, yet they depend on the knowledge base they are applied to. Thus, also the choice of the optimal feature sets deserves a preliminary feature construction phase. We perform this pre-processing step by means of a randomized search procedure based on *simulated annealing*. The reason of such a choice is to avoid the traversal of the almost entire (potentially large) search space.

Early clustering approaches devised for terminological representations pursued a way for attacking the problem with logic-based methods for specific DLs [19, 13]. However, it has been pointed out that these methods may suffer from noise in the data. This motivates our investigation on similarity-based clustering methods which can be more noise-tolerant, and as language-independent as possible. Specifically, we propose a multi-relational extension of effective clustering techniques, namely the (BISECTING) K-MEANS approach [16] (originally developed for numeric or ordinal features) that is tailored for the SW context. Instead of the notion of *means* that characterizes the algorithms descending from (BISECTING) K-MEANS approach, we recur to the notion of *medoids* [18] that represent the central individual of a cluster, namely the one having the minimum dissimilarity value w.r.t. the other individuals in the same cluster. Hence we propose a sort of BISECTING AROUND MEDOIDS algorithm, which exploits the aforementioned fully semantic and language independent measures to work on DLs.

An initial evaluation of the method applied to real ontologies is presented. It is based on the measurement of internal validity indices such as the silhouette measure [18].

The paper is organized as follows. Sect. 2 recalls the basics of the representation and the distance measure adopted. The clustering algorithm is presented and discussed in Sect. 3. The experimentation with OWL ontologies proving the validity of the approach is presented in Sect. 5. After Sect. 4 concerning the related work, conclusions are finally examined in Sect. 6.

2   Semantic Distance Measures

In the following, we assume that resources, concepts and their relationship may be defined in terms of a generic ontology language that may be mapped to some DL language with the standard model-theoretic semantics (see the handbook [1] for a thorough reference). In this context, a *knowledge base* $\mathcal{K} = \langle \mathcal{T}, \mathcal{A} \rangle$ contains a *TBox* $\mathcal{T}$ and an *ABox* $\mathcal{A}$. $\mathcal{T}$ is a set of concept definitions. $\mathcal{A}$ contains assertions (facts, data) concerning the world state. Normally, the *unique names assumption* is made on the ABox individuals[1] therein. The set of the individuals occurring in $\mathcal{A}$ will be denoted with $\mathsf{Ind}(\mathcal{A})$.

Differently from reasoning with databases (and logic programs), for which the *Closed World Assumption* (CWA) is natural, these knowledge bases are to be regarded as inherently incomplete, as a consequence of the *Open World Assumption* (OWA) that is made on the semantics. This makes then suitable for the SW applications that foresee the continuous increase of available resources across the Web.

As regards the inference services, like all other instance-based methods, our procedure may require performing *instance-checking*, which amounts to determining whether an individual, say a, belongs to a concept extension, i.e. whether $C(a)$ holds for a certain concept C.

2.1 A Semantic Semi-distance for Individuals

For our purposes, a function for measuring the similarity of individuals rather than concepts is needed. However, differently from the concepts in DLs, individuals do not have a syntactic structure that can be compared. This has led to lifting them to the concept description level before comparing them (recurring to the approximation of the *most specific concept*[2] of an individual w.r.t. the ABox).

We have developed a new measure for assessing the similarity of individuals in a knowledge base that does not depends on the notion of *most specific concept* and whose definition totally depends on semantic aspects of the individuals. It is grounded on the intuition that, on a semantic level, similar individuals should behave similarly with respect to the same concepts that is, similar individuals should be instances of almost the same concepts. Following the ideas borrowed from ILP [27] and *propositionalization* methods [21], the measure assess the similarity value by comparing the semantics of individuals along a number of dimensions represented by a *committee* of concept descriptions. Specifically, the rationale of the new measure is to compare them on the grounds of their behavior w.r.t. a given set of hypotheses, that is a collection of concept descriptions, say $\mathsf{F} = \{F_1, F_2, \ldots, F_m\}$, which stands as a group of discriminating *features* expressed in the language taken into account.

In its simple formulation, a family of distance functions for individuals inspired to Minkowski's distances can be defined as follows:

---

[1] Individuals can be assumed to be identified by their own URI.

[2] The most specific concept of an individual a w.r.t. the ABox is the concept that is most specific (w.r.t. the subsuption relationship) of which a is instance (see [1] for more details).

Definition 1 (family of measures). *Let* $\mathcal{K} = \langle \mathcal{T}, \mathcal{A} \rangle$ *be a knowledge base. Given set of concept descriptions* $\mathsf{F} = \{F_1, F_2, \ldots, F_m\}$, *a family of functions*

$$d_p^\mathsf{F} : \mathsf{Ind}(\mathcal{A}) \times \mathsf{Ind}(\mathcal{A}) \mapsto \mathbb{R}$$

is defined as follows:

$$\forall a, b \in \mathsf{Ind}(\mathcal{A}) \quad d_p^\mathsf{F}(a, b) := \frac{1}{m} \left(\sum_{i=1}^{m} \mid \pi_i(a) - \pi_i(b) \mid^p \right)^{1/p}$$

where $p > 0$ *and* $\forall i \in \{1, \ldots, m\}$ *the* projection functions π_i *are defined:*

$$\forall a \in \mathsf{Ind}(\mathcal{A}) \quad \pi_i(a) = \begin{cases} 1 & F_i(x) \in \mathcal{A} \\ 0 & \neg F_i(x) \in \mathcal{A} \\ 1/2 & otherwise \end{cases}$$

The superscript F will be omitted when the set of hypotheses is fixed.

As an alternative, the definition of the measures can be made more accurate by considering entailment rather than the simple ABox look-up, when determining the values of the projection functions:

$$\forall a \in \mathsf{Ind}(\mathcal{A}) \quad \pi_i(a) = \begin{cases} 1 & \mathcal{K} \models F_i(a) \\ 0 & \mathcal{K} \models \neg F_i(a) \\ 1/2 & otherwise \end{cases}$$

It is easy to prove that these functions have the standard properties for semi-distances (also called semi-metrics) [4]:

Proposition 1 (semi-distance). *For a fixed feature set and* $p > 0$, *given any three instances* $a, b, c \in \mathsf{Ind}(\mathcal{A})$. *it holds that:*

1. $d_p(a, b) > 0$
2. $d_p(a, b) = d_p(b, a)$
3. $d_p(a, c) \leq d_p(a, b) + d_p(b, c)$

Proof. 1. and 2. are trivial. As for 3., noted that

$$(d_p(a, c))^p = \frac{1}{m^p} \sum_{i=1}^{m} \mid \pi_i(a) - \pi_i(c) \mid^p = \frac{1}{m^p} \sum_{i=1}^{m} \mid \pi_i(a) - \pi_i(b) + \pi_i(b) - \pi_i(c) \mid^p$$

$$\leq \frac{1}{m^p} \sum_{i=1}^{m} \mid \pi_i(a) - \pi_i(b) \mid^p + \frac{1}{m^p} \sum_{i=1}^{m} \mid \pi_i(b) - \pi_i(c) \mid^p$$

$$\leq (d_p(a, b))^p + (d_p(b, c))^p \leq (d_p(a, b) + d_p(b, c))^p$$

then the property follows for the monotonicity of the power function.

It cannot be proved that $d_p(a, b) = 0$ iff $a = b$. This is the case of *indiscernible* individuals with respect to the given set of hypotheses F. This does not cause unexpected behavior if the committee of features is really discriminating for the individuals.

Compared to other proposed distance (or dissimilarity) measures [6], the presented function does not depend on the constructors of a specific language, rather it requires only retrieval or instance-checking service used for deciding whether an individual belongs to a concept extension.

Note that the π_i functions ($\forall i = 1, \ldots, m$) for the training instances, that contribute to determine the measure with respect to new ones, can be computed in advance thus determining a speed-up in the actual computation of the measure. This is very important for the measure integration in algorithms which massively use this distance, such as all instance-based methods.

The underlying idea for the measure is that similar individuals should exhibit the same behavior w.r.t. the concepts in F. Here, we make the assumption that the feature-set F represents a sufficient number of (possibly redundant) features that are able to really discriminate different individuals.

2.2 Feature Set Optimization

Experimentally, we could obtain good results by using the very set of both primitive and defined concepts found in the ontology (see Sect. 5). However for some knowledge bases these concepts may be insufficient to guarantee the discernibility of different individuals. Moreover also the unlikely case of knowledge bases containing many roles and few concepts must be taken into account.

Therefore, the choice of the concepts to be included – *feature selection* – may be crucial, for a good committee may discern the individuals better and a possibly smaller committee yields more efficiency when computing the distance.

We have devised a specific optimization algorithms founded in *genetic programming* [11] and *simulated annealing* which are able to find optimal choices of discriminating concept committees. Namely, since the function is very dependent on the concepts included in the committee of features F, two immediate heuristics can be derived: 1) control the number of concepts of the committee, including especially those that are endowed with a real discriminating power; 2) finding optimal sets of discriminating features, by allowing also their composition employing the specific constructors made available by the representation language of choice.

Both these objectives can be accomplished by means of machine learning techniques especially when knowledge bases with large sets of individuals are available. Namely, part of the entire data can be hold out in order to learn optimal F sets, in advance with respect to the successive usage for all other purposes.

We have also experimented the usage of genetic programming for constructing an optimal set of features [11]. Yet these methods may suffer from being possibly caught in local minima. An alternative is employing a different probabilistic search procedure which aims at a global optimization, such as taboo search or random walk algorithms. For simplicity, we devised a simulated annealing search, whose algorithm is depicted in Fig. 1.

FeatureSet OPTIMIZEFEATURESET($\mathcal{K}$, ΔT)
input $\mathcal{K}$: Knowledge base
         ΔT: function controlling the decrease of temperature
output FeatureSet
static   currentFS: current Feature Set
         nextFS: next Feature Set
         Temperature: controlling the probability of downward steps
begin
currentFS ← MAKEINITIALFS($\mathcal{K}$);
for $t ← 1$ **to** ∞ **do**
        Temperature ← Temperature − $\Delta T(t)$;
        if (Temperature $= 0$)
            return currentFS;
        nextFS ← RANDOMSUCCESSOR(currentFS,$\mathcal{K}$);
        ΔE ← FITNESSVALUE(nextFS) − FITNESSVALUE(currentFS);
        if ($\Delta E > 0$)
            currentFS ← nextFS;
        else   // *replace FS with given probability*
            REPLACE(currentFS, nextFS, $e^{\Delta E/\text{Temperature}}$);
end

Fig. 1 Feature set optimization procedure based on *simulated annealing*

The algorithm searches the space of all possible feature committees starting from an initial guess (determined by MAKEINITIALFS($\mathcal{K}$)) based on the concepts (both primitive and defined) currently referenced in the knowledge base. The loop controlling the search is repeated for a number of times that depends on the temperature which gradually decays to 0, when the current committee can be returned. The current feature set is iteratively refined calling a suitable procedure RANDOMSUCCESSOR(). Then the fitness of the new feature set is compared to that of the previous one determining the increment of energy ΔE. If this is non-null then the computed committee replaces the old one. Otherwise it will be replaced with a probability that depends on ΔE.

As regards the FITNESSVALUE(F), it can be computed as the *discernibility factor* of the feature set. For example given the whole set of individuals $IS = \mathsf{Ind}(\mathcal{A})$ (or just a sample to be used to induce an optimal measure) the fitness function may be:

$$\text{FITNESSVALUE}(\mathsf{F}) = \sum_{(a,b)\in IS^2} \sum_{i=1}^{|\mathsf{F}|} |\ \pi_i(a) - \pi_i(b)\ |$$

As concerns finding candidates to replace the current committee (RANDOMSUCCESSOR()), the function was implemented by recurring to simple transformations of a feature set:

- adding (resp. removing) a concept C: nextFS $\leftarrow$ currentFS $\cup \{C\}$
  (resp. nextFS $\leftarrow$ currentFS $\setminus \{C\}$)
- randomly choosing one of the current concepts from currentFS, say C;
  replacing it with one of its refinements $C' \in \text{REF}(C)$

Refinement of concept description is language specific. E.g. for the case of $\mathcal{ALC}$ logic, refinement operators have been proposed in [15, 22]. An alternative optimization procedure based on genetic programming has been also proposed [11].

3   Grouping Individuals by Hierarchical Clustering

The conceptual clustering procedure that we propose can be ascribed to the category of the heuristic partitioning algorithms such as K-MEANS and EM [16]. For the categorical nature of the assertions on individuals, the notion of mean is replaced by the one of medoid, as in PAM (*Partition Around Medoids* [18]). Besides it is crafted to work iteratively to produce a hierarchical clustering.

The algorithm implements a top-down bisecting method starting with one universal cluster grouping all instances. Iteratively, it creates two new clusters by bisecting an existing one and this continues until the desired number of clusters is reached. This algorithm can be thought as levelwise producing a dendrogram: the number of levels coincides with the number of clusters.

Each cluster is represented by one of its individuals. We consider the notion of medoid as representing a cluster center, since our distance measure works on a categorical feature-space. The medoid of a group of individuals is the individual that has the lowest distance w.r.t. the others. Formally. given a cluster $C = \{a_1, a_2, \ldots, a_n\}$, the medoid is defined as:

$$m = \text{medoid}(C) = \operatorname*{argmin}_{a \in C} \sum_{j=1}^{n} d(a, a_j)$$

The proposed method can be considered as a hierarchical extension of PAM. A bi-partition is repeated level-wise producing a dendrogram. Fig. 2 reports a sketch of our algorithm. It essentially consists of two nested loops: the outer one computes a new level of the resulting dendrogram and it is repeated until the desired number of clusters is obtained; the inner loop consists of a run of the PAM algorithm at the current level.

Per each level, the next worst cluster is selected (SELECTWORSTCLUS-TER() function) on the grounds of its quality, e.g. the one endowed with the least average inner similarity (or cohesiveness [28]). This cluster is candidate to being splitted. The partition is constructed around two medoids initially chosen (SELECTMOSTDISSIMILAR() function) as the most dissimilar elements in the cluster and then iteratively adjusted in the inner loop. In the end, the candidate cluster is replaced by the newly found parts at the next level of the dendrogram.

```
clusterVector HIERARCHICALBISECTINGAROUNDMEDOIDS(allIndividuals, k, maxIterations)
input    allIndividuals: set of individuals
         k: number of clusters;
         maxIterations: max number of inner iterations;
output clusterVector: array [1..k] of sets of clusters

begin
level ← 0;
clusterVector[1] ← allIndividuals;
repeat
         ++level;
         cluster2split ← SELECTWORSTCLUSTER(clusterVector[level]);
         iterCount ← 0;
         stableConfiguration ← false;
         (newMedoid1,newMedoid2) ← SELECTMOSTDISSIMILAR(cluster2split);
         repeat
            ++iterCount;
            (medoid1,medoid2) ← (newMedoid1,newMedoid2);
            (cluster1,cluster2) ← DISTRIBUTE(cluster2split,medoid1,medoid2);
            newMedoid1 ← MEDOID(cluster1);
            newMedoid2 ← MEDOID(cluster2);
            stableConfiguration ← (medoid1 = newMedoid1) ∧ (medoid2 = newMedoid2);
         until stableConfiguration ∨ (iterCount = maxIterations);
         clusterVector[level+1] ← REPLACE(cluster2split,cluster1,cluster2,clusterVector[level]);
until (level = k);
end
```

Fig. 2 The HIERARCHICAL BISECTING AROUND MEDOIDS Algorithm

The inner loop basically resembles to a 2-MEANS algorithm, where medoids are considered instead of means that can hardly be defined in symbolic computations. Then, the standard two steps are performed iteratively:

distribution   given the current medoids, distribute the other individuals to either partition on the grounds of their distance w.r.t. the respective medoid;

center re-computation   given the bipartition obtained by DISTRIBUTE(), compute the new medoids for either cluster.

The medoid tend to change at each iteration until eventually they converge to a stable couple (or when a maximum number of iterations have been performed).

3.1 From Clusters to Concepts

Each node of the tree (a cluster) may be labeled with an intensional concept definition which characterizes the individuals in the given cluster while discriminating those in the twin cluster at the same level. Labeling the tree-nodes with concepts can be regarded as solving a number of supervised learning problems in the specific multi-relational representation targeted in our setting. As such it deserves specific solutions that are suitable for the DL languages employed.

A straightforward solution may be found, for DLs that allow for the computation of (an approximation of) the *most specific concept* (msc) and *least common subsumer* (lcs) [1], such as $\mathcal{ALN}$, $\mathcal{ALE}$ or $\mathcal{ALC}$. This may involve the following steps:

given a cluster of individuals node_j

- **for each** individual $a_i \in \mathsf{node}_j$ **do**

    compute $M_i \leftarrow \mathsf{msc}(a_i)$ w.r.t. $\mathcal{A}$;

- **let** $\mathsf{MSCs}_j \leftarrow \{M_i \mid a_i \in \mathsf{node}_j\}$;
- **return** $\mathsf{lcs}(\mathsf{MSCs}_j)$

As an alternative, algorithms for learning concept descriptions expressed in DLs may be employed, such as YINYANG [15] or DL-LEARNER [23]. Indeed, concept formation can be cast as a supervised learning problem: once the two clusters at a certain level have been found, where the members of a cluster are considered as positive examples and the members of the dual cluster as negative ones. Then, in principle, any concept learning method which can deal with these representations may be utilized for this new task.

3.2 *Discussion*

An adaptation of a PAM algorithm has several favorable properties. Since it performs clustering with respect to any specified metric, it allows a flexible definition of similarity. This flexibility is particularly important in biological applications where researchers may be interested, for example, in grouping correlated or possibly also anti-correlated elements. Many clustering algorithms do not allow for a flexible definition of similarity, but allow only Euclidean distance in current implementations.

In addition to allowing a flexible distance metric, a PAM algorithm has the advantage of identifying clusters by the medoids. Medoids are robust representations of the cluster centers that are less sensitive to outliers than other cluster profiles, such as the cluster means of K-MEANS. This robustness is particularly important in the common context that many elements do not belong exactly to any cluster, which may be the case of the membership in DL knowledge bases, which may be not ascertained given the OWA.

The representation of centers by means of medoids has two advantages. First, it presents no limitations on attributes types, and, second, the choice of medoids is dictated by the location of a predominant fraction of points inside a cluster and, therefore, it is lesser sensitive to the presence of outliers. In

K-MEANS case a cluster is represented by its centroid, which is a mean (usually weighted average) of points within a cluster. This works conveniently only with numerical attributes and can be negatively affected by a single outlier.

4   Related Work

As previously mentioned, the application of distance based methods to unsupervised learning with knowledge bases represented in DLs is relatively new, except for some pioneering work. To the best of our knowledge there are very few examples of similar clustering algorithms working on complex representations that are suitable for knowledge bases of semantically annotated resources.

The unsupervised learning procedure presented in this paper is mainly based on two factors: the semantic dissimilarity measures and the clustering method. Thus, in this section, we briefly discuss sources of inspiration for our procedure and some related approaches.

4.1   Relational Similarity Measures

As previously mentioned, various attempts to define semantic similarity (or dissimilarity) measures for concept languages have been made, yet they have still a limited applicability to simple languages [6] or they are not completely semantic depending also on the structure of the descriptions [7]. Very few works deal with the comparison of individuals rather than concepts.

In the context of clausal logics, a metric was defined [25] for the Herbrand interpretations of logic clauses as induced from a distance defined on the space of ground atoms. This kind of measures may be employed to assess similarity in *deductive databases*. Although it represents a form of fully semantic measure, different assumptions are made with respect to those which are standard for knowledgeable bases in the SW perspective. Therefore the transposition to the context of interest is not straightforward.

Our measure is mainly based on Minkowski's measures [29] and on a method for distance induction developed by Sebag [27] in the context of *machine learning*, where *metric learning* is developing as an important subfield. In this work it is shown that the induced measure could be accurate when employed for classification tasks even though set of features to be used were not the optimal ones (or they were redundant). Indeed, differently from our unsupervised learning approach, the original method learns different versions of the same target concept, which are then employed in a voting procedure similar to the Nearest Neighbor approach for determining the classification of instances.

A source of inspiration was also *rough sets* theory [26] which aims at the formal definition of vague sets by means of their approximations determined by an indiscernibility relationship. Hopefully, these methods developed in this context will help solving the open points of our framework (see Sect. 6) and suggest new ways to treat uncertainty.

4.2 Clustering Procedures

Our algorithm adapts to the specific representations devised for the SW context a combination of the distance-based approaches (see [16]). Specifically, in the methods derived from K-MEANS and K-MEDOIDS each cluster is represented by one of its points.

PAM and CLARA [18] are early systems adopting this approach. They implement iterative optimization methods that essentially cyclically relocate points between perspective clusters and recompute potential medoids.

Further comparable clustering methods are those based on an *indiscernibility relationship* [14]. While in our method this idea is embedded in the semi-distance measure (and the choice of the committee of concepts), these algorithms are based on an iterative refinement of an equivalence relationship which eventually induces clusters as equivalence classes.

Alternatively evolutionary clustering approaches may be considered [11] which are also capable to determine a good estimate of the number of clusters. The UNC algorithm is a more recent related approach which was also extended to the hierarchical clustering case H-UNC [24].

As mentioned in the introduction, the classic approaches to conceptual clustering [28] in complex (multi-relational) spaces are based on structure and logics. Kietz & Morik proposed a method for efficient construction of knowledge bases for the BACK concept language [19]. This method exploits the assertions concerning the roles available in the knowledge base, in order to assess, in the corresponding relationship, those subgroups of the domain and ranges which may be inductively deemed as disjoint. In the successive phase, supervised learning methods are used on the discovered disjoint subgroups to construct new concepts that account for them. A similar approach is followed in [13], where the supervised phase is performed as an iterative refinement step, exploiting suitable refinement operators for a different DL, namely $\mathcal{ALC}$.

5 Experimental Validation

An experimental session was planned in order to prove the clustering method feasible. It could not be a comparative experimentation since, to the best of our knowledge no other hierarchical clustering method has been proposed which is able to cope with DLs representations (on a semantic level) except [19, 13] which are language-dependent and produce non-hierarchical clusterings.

5.1 Settings

A number of different ontologies represented in OWL were selected for the experiments, namely: FSM, SURFACE-WATER-MODEL (SWM), TRANSPORTATION and NEWTESTAMENTNAMES (NTN) from the Protégé library[3], the FINANCIAL ontology[4] employed as a testbed for the PELLET reasoner[5]. Table 1 summarizes important details concerning the ontologies employed in the experimentation. For each individual, a variable number of assertions was available in the KB, hence the number of triples in each ontology is in an order of magnitude of tens of thousands.

Table 1 Ontologies employed in the experiments

ontology	DL	#concepts	#obj. prop.	#data prop.	#individuals
FSM	$\mathcal{SOF}(D)$	20	10	7	37
S.-W.-M.	$\mathcal{ALCOF}(D)$	19	9	1	115
TRANSPORTATION	$\mathcal{ALC}$	44	7	0	250
FINANCIAL	$\mathcal{ALCIF}$	60	17	0	652
NTN	$\mathcal{SHIF}(D)$	47	27	8	676

As pointed out in several surveys on clustering, it is better to use a different criterion for clustering (e.g. for choosing the candidate cluster to bisection) and for assessing the quality of a cluster. In the evaluation we employed standard validity measures for clustering adapted to the categorical nature of our application: the *within-cluster sum of square errors* (WSS), a measure of cohesion and the classic *silhouette* measure [18] both substituting the role of cluster center with the cluster medoid.

Besides, we propose also the extension of Dunn's validity index [3] for clusterings produced by the algorithm. Namely, we define a modified version of this index to deal with medoids. Let $P = \{C_1, \ldots, C_k\}$ be a possible clustering of n individuals in k clusters. The new index can be defined:

$$V_{GD}(P) = \min_{1 \le i \le k} \left\{ \min_{1 \le j \le k,\ i \ne j} \left\{ \delta_p(C_i, C_j) / \max_{1 \le h \le k} \{\Delta_p(C_h)\} \right\} \right\}$$

where δ_p is the Hausdorff distance for clusters, derived from d_p:

$$\delta_p(C_i, C_j) = (1/|C_i| + |C_j|)(\textstyle\sum_{a \in C_i} d_p(a, \mathrm{med}(C_i)) + \sum_{b \in C_j} d_p(b, \mathrm{med}(C_j)))$$

and the cluster diameter measure Δ_p can be defined in many ways [3]; we chose the definition: $\Delta_p(C_h) = (2/|C_h|) \sum_{c \in C_h} d_p(c, \mathrm{med}(C_h))$ which is more noise-tolerant w.r.t. other proposed measures.

---

[3] http://protege.stanford.edu/plugins/owl/owl-library

[4] http://www.cs.put.poznan.pl/alawrynowicz/financial.owl

[5] http://pellet.owldl.com

For each populated ontology, the experiments have been repeated for varying numbers k of final clusters (5 through 25). In the computation of the distances between individuals (the most time-consuming operation) all concepts in the ontology have been used for the committee of features, thus guaranteeing meaningful measures with high redundancy. The PELLET reasoner[6] was employed to compute the projection functions values for each individual that are needed to compute the metric. An overall experimentation of 16 repetitions on a single dataset took a maximum of 40 minutes on a 2.0GhZ (2GB RAM) Core2Duo Linux box.

5.2   Outcomes

In this exploratory analysis of an unsupervised method, the aim is not determining an absolute performance that can be compared to other methods. Rather, using also the range of values of the indices, one may detect the choices of k (number of clusters to be produced) that are better suited for the datasets. Moreover, the variability of the index values may also used to judge the stability of the method.

The outcomes of the experiments are reported in Fig. 3. For each ontology, the graphs for Dunn's, silhouette and WSS indices, respectively, are depicted for increasing values of the input parameter k.

Preliminarily, it is possible to note that the silhouette values are quite stably close to the top of the range (+1). This may also give a way to assess the absolute effectiveness of our algorithms w.r.t. others, although applied to different representations.

The modified Dunn's index decreases with higher values of k also as a consequence of the definition in terms of medoids instead of centers. However, the measure presents several plateaus where limited variations are observed and abrupt descents at some levels.

Conversely, the cohesion coefficient WSS exhibits more variability, indicating that, for some levels, the clustering found by the algorithm is not the natural one. However it must be taken into account the meaning of this index which measures the average cohesion of the clusters, while the others measure a balance of intra-cluster cohesion and inter-clusters separation.

Hence, from an exploratory point of view, the values of the three indices may hint possible cut points (the *knees* in the curves) in the hierarchical clusterings (i.e. optimal values of k for stopping the partitional process).

These outcomes may be partially compared to those resulting from an evolutionary clustering method (based on the same pseudo-metrics) which directly produces k clusters [11]. Indeed some of the datasets and validity indices used there were also employed in this experimentation.

---

[6] Version 1.5.1.

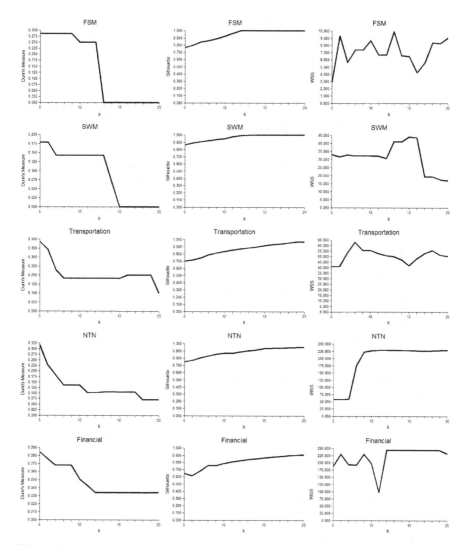

Fig. 3 Outcomes of the experiments: Dunn's, Silhouette, and WSS index graphs

6 Conclusions

This work has presented a clustering method that can be applied to (multi-) relational representations which are standard for SW applications. Namely, it can be used to discover interesting hierarchical groupings of semantically annotated resources in a wide range of concept languages. The algorithm, is an adaptation of the classic BISECTING K-MEANS to complex representations typical of the ontology in the SW.

The method exploits a dissimilarity measure, that is based on the knowledge base semantics w.r.t. a number of dimensions corresponding to a committee of discriminating features represented by a set of concept descriptions. We have also proposed a procedure based on stochastic search to induce optimal feature sets when the concepts defined in the knowledge base are not able to tell the difference between different individuals. Moreover this may be applied also in the unlikely case of knelled bases containing many roles and few concepts.

Currently we are investigating (flat) evolutionary clustering methods both for performing the optimization of the feature committee and for clustering individuals, which are also able to automatically discover an optimal number of clusters [11].

References

1. Baader, F., Calvanese, D., McGuinness, D., Nardi, D., Patel-Schneider, P. (eds.): The Description Logic Handbook. Cambridge University Press, Cambridge (2003)
2. Berners-Lee, T., Hendler, J., Lassila, O.: The Semantic Web. Scientific American 284(5), 34–43 (2001)
3. Bezdek, J.C., Pal, N.R.: Some new indexes of cluster validity. IEEE Transactions on Systems, Man, and Cybernetics 28(3), 301–315 (1998)
4. Bock, H.H., Diday, E.: Analysis of symbolic data: exploratory methods for extracting statistical information from complex data. Springer, Heidelberg (2000)
5. Borgida, A.: On the relative expressiveness of description logics and predicate logics. Artificial Intelligence 82(1-2), 353–367 (1996)
6. Borgida, A., Walsh, T.J., Hirsh, H.: Towards measuring similarity in description logics. In: Horrocks, I., Sattler, U., Wolter, F. (eds.) Working Notes of the International Description Logics Workshop, CEUR Workshop Proceedings, Edinburgh, UK, vol. 147 (2005)
7. d'Amato, C., Fanizzi, N., Esposito, F.: Reasoning by analogy in description logics through instance-based learning. In: Tummarello, G., Bouquet, P., Signore, O. (eds.) Proceedings of Semantic Web Applications and Perspectives, 3rd Italian Semantic Web Workshop, SWAP 2006. CEUR Workshop Proceedings, Pisa, Italy, vol. 201 (2006)
8. d'Amato, C., Staab, S., Fanizzi, N., Esposito, F.: Efficient discovery of services specified in description logics languages. In: Di Noia, T., et al. (eds.) Proceedings of the SMR^2 2007 Workshop on Service Matchmaking and Resource Retrieval in the Semantic Web (SMRR 2007) co-located with ISWC 2007 + ASWC 2007, CEUR Workshop Proceedings. Busan, South Korea, vol. 243. CEUR (2007)
9. Dean, M., Schreiber, G.: Web Ontology Language Reference. W3C recommendation, W3C (2004), http://www.w3.org/TR/owl-ref
10. Emde, W., Wettschereck, D.: Relational instance-based learning. In: Saitta, L. (ed.) Proceedings of the 13th International Conference on Machine Learning, ICML 1996, pp. 122–130. Morgan Kaufmann, San Francisco (1996)

11. Fanizzi, N., d'Amato, C., Esposito, F.: Randomized metric induction and evolutionary conceptual clustering for semantic knowledge bases. In: Silva, M., Laender, A., Baeza-Yates, R., McGuinness, D., Olsen, O., Olstad, B. (eds.) Proceedings of the ACM International Conference on Knowledge Management, CIKM 2007. Lisbon Portugal. ACM Press, New York (2007)

12. Fanizzi, N., d'Amato, C., Esposito, F.: Conceptual clustering and its application to concept drift and novelty detection. In: Bechhofer, S., Hauswirth, M., Hoffmann, J., Koubarakis, M. (eds.) ESWC 2008. LNCS, vol. 5021, pp. 318–332. Springer, Heidelberg (2008)

13. Fanizzi, N., Iannone, L., Palmisano, I., Semeraro, G.: Concept formation in expressive description logics. In: Boulicaut, J.-F., Esposito, F., Giannotti, F., Pedreschi, D. (eds.) ECML 2004. LNCS (LNAI), vol. 3201, pp. 99–110. Springer, Heidelberg (2004)

14. Hirano, S., Tsumoto, S.: An indiscernibility-based clustering method. In: Hu, X., Liu, Q., Skowron, A., Lin, T.Y., Yager, R., Zhang, B. (eds.) 2005 IEEE International Conference on Granular Computing, pp. 468–473. IEEE Computer Society Press, Los Alamitos (2005)

15. Iannone, L., Palmisano, I., Fanizzi, N.: An algorithm based on counterfactuals for concept learning in the semantic web. Applied Intelligence 26(2), 139–159 (2007)

16. Jain, A.K., Murty, M.N., Flynn, P.J.: Data clustering: A review. ACM Computing Surveys 31(3), 264–323 (1999)

17. Janowicz, K.: Sim-dl: Towards a semantic similarity measurement theory for the description logic $\mathcal{ALCNR}$ in geographic information retrieval. In: Meersman, R., Tari, Z., Herrero, P. (eds.) OTM 2006 Workshops. LNCS, vol. 4278, pp. 1681–1692. Springer, Heidelberg (2006)

18. Kaufman, L., Rousseeuw, P.: Finding Groups in Data: an Introduction to Cluster Analysis. John Wiley & Sons, Chichester (1990)

19. Kietz, J.U., Morik, K.: A polynomial approach to the constructive induction of structural knowledge. Machine Learning 14(2), 193–218 (1994)

20. Kirsten, M., Wrobel, S.: Relational distance-based clustering. In: Page, D.L. (ed.) ILP 1998. LNCS (LNAI), vol. 1446, pp. 261–270. Springer, Heidelberg (1998)

21. Kramer, S., Lavrač, N., Džeroski, S.: Propositionalization approaches to relational data mining. In: Džeroski, S., Lavrač, N. (eds.) Relational Data Mining, Springer, Heidelberg (2001)

22. Lehmann, J., Hitzler, P.: Foundations of refinement operators for description logics. In: Blockeel, H., Ramon, J., Shavlik, J., Tadepalli, P. (eds.) ILP 2007. LNCS (LNAI), vol. 4894, pp. 161–174. Springer, Heidelberg (2008)

23. Lehmann, J., Hitzler, P.: A refinement operator based learning algorithm for the $\mathcal{ALC}$ description logic. In: Blockeel, H., Ramon, J., Shavlik, J., Tadepalli, P. (eds.) ILP 2007. LNCS, vol. 4894, pp. 147–160. Springer, Heidelberg (2008)

24. Nasraoui, O., Krishnapuram, R.: One step evolutionary mining of context sensitive associations and web navigation patterns. In: Proceedings of the SIAM conference on Data Mining, Arlington, VA, pp. 531–547 (2002)

25. Nienhuys-Cheng, S.H.: Distances and limits on Herbrand interpretations. In: Page, D.L. (ed.) ILP 1998. LNCS (LNAI), vol. 1446, pp. 250–260. Springer, Heidelberg (1998)

26. Pawlak, Z.: Rough Sets: Theoretical Aspects of Reasoning About Data. Kluwer Academic Publishers, Dordrecht (1991)
27. Sebag, M.: Distance induction in first order logic. In: Džeroski, S., Lavrač, N. (eds.) ILP 1997. LNCS (LNAI), vol. 1297, pp. 264–272. Springer, Heidelberg (1997)
28. Stepp, R.E., Michalski, R.S.: Conceptual clustering of structured objects: A goal-oriented approach. Artificial Intelligence 28(1), 43–69 (1986)
29. Zezula, P., Amato, G., Dohnal, V., Batko, M.: Similarity Search – The Metric Space Approach. In: Advances in Database Systems, Springer, Heidelberg (2007)

The *Ex* Project: Web Information Extraction Using Extraction Ontologies

Martin Labský, Vojtěch Svátek, Marek Nekvasil, and Dušan Rak

Abstract. Extraction ontologies represent a novel paradigm in web information extraction (as one of 'deductive' species of web mining) allowing to swiftly proceed from initial domain modelling to running a functional prototype, without the necessity of collecting and labelling large amounts of training examples. Bottlenecks in this approach are however the tedium of developing an extraction ontology adequately covering the semantic scope of web data to be processed and the difficulty of combining the ontology-based approach with inductive or wrapper-based approaches. We report on an ongoing project aiming at developing a web information extraction tool based on richly-structured extraction ontologies and with additional possibility of (1) semi-automatically constructing these from third-party domain ontologies, (2) absorbing the results of inductive learning for subtasks where pre-labelled data abound, and (3) actively exploiting formatting regularities in the wrapper style.

1 Introduction

Web information extraction (WIE) represents a specific category of web mining. It consists in the identification of typically small pieces of relevant text within web pages and their aggregation into larger structures such as data records or instances of ontology classes. As its core task is application of pre-existent patterns or models (in contrast to inductively discovering new patterns), it falls under the notion of 'deductive' web mining [15], similarly as e.g. web document classification. As such, some kind of prior knowledge is indispensable in WIE. However, the 'deductive' aspects of WIE are often complemented with inductive ones, especially in terms of learning the patterns/models (at least partly) from training data.

Martin Labský, Vojtěch Svátek, Marek Nekvasil, and Dušan Rak
Dept. of Information and Knowledge Engineering, University of Economics, Prague,
W. Churchill Sq. 4, 130 67 Praha 3, Czech Republic
e-mail: {labsky,svatek,nekvasim,rakdusan}@vse.cz

B. Berendt et al. (Eds.): Knowl. Disc. Enhan. with Sem. and Soc. Info., SCI 220, pp. 71–88.
springerlink.com © Springer-Verlag Berlin Heidelberg 2009

In the last decade, WIE was actually dominated by two paradigms. One—*wrapper-based*—consists in systematically exploiting the surface structure of HTML code, assuming the presence of regular structures that can be used as anchors for the extraction. This approach is now widely adopted in industry, however, its dependence on formatting regularity limits its use for diverse categories of web pages. The other—*inductive*—paradigm assumes the presence of training data: either web pages containing pre-annotated tokens or stand-alone examples of data instances. It is linked to exploration of various computational learning paradigms, e.g. Hidden-Markov Models, Maximum Entropy Models, Conditional Random Fields [9] or symbolic approaches such as rule learning [1]. Again, however, the presence of (sufficient amounts of) annotated training data is a pre-condition that is rarely fulfilled in real-world settings, and manual labelling of training data is often unfeasible; statistical bootstrapping alleviates this problem to some degree but at the same time it may introduce complexity and side-effects not transparent to a casual user of a WIE tool. In addition, both approaches usually deliver extracted information as rather weakly semantically structured; if WIE is to be used to fuel semantic web repositories, secondary mapping to ontologies is typically needed, which makes the process complicated and possibly error-prone.

There were recently proposals for pushing ontologies towards the actual extraction process as immediate prior knowledge.[1] *Extraction ontologies*[2] [3] define the concepts, the instances of which are to be extracted, in the sense of various attributes, their allowed values as well as higher level (e.g. cardinality or mutual dependency) constraints. Extraction ontologies are assumed to be hand-crafted based on observation of a sample of resources; however, due to their clean and rich conceptual structure (allowing partial intra-domain reuse and providing immediate semantics to extracted data), they are superior to ad-hoc hand-crafted patterns used in early times of WIE. At the same time, they allow for rapid start of the actual extraction process, as even a very simple extraction ontology (designed by a competent person) is likely to cover a sensible part of target data and generate meaningful feedback for its own redesign; several iterations are of course needed to obtain results in sufficient quality. It seems that for web domains that consist of a high number of relatively tiny and evolving resources (such as web product catalogs), information extraction ontologies are the first choice. However, to make maximal use of available data and knowledge and avoid overfitting to a few data resources examined by the designer, the whole process must not neglect available labelled data, formatting regularities and even pre-existing domain ontologies.

In this paper we report on an ongoing effort in building a WIE tool named *Ex*, which would synergistically exploit all the mentioned resources, with central role of extraction ontologies. The paper is structured as follows. Section 2 explains the structure of extraction ontologies used in *Ex*. Section 3 describes the steps of the information extraction process using extraction ontologies and other resources. Section 4 briefly reports on experiments in three different domains. Finally, section 5 surveys related research, and section 6 outlines future work.

---

[1] See [8] for general discussion of the types and roles of ontologies in the WIE process.

[2] In earlier work [7] we used the term *presentation ontology*.

2 Ex(traction) Ontology Content

Extraction ontologies in *Ex* are designed so as to extract occurrences of *attributes* (such as 'age' or 'surname'), i.e. standalone named entities or values, and occurrences of whole *instances* of *classes* (such as 'person'), as groups of attributes that 'belong together', from HTML pages (or texts in general) in a domain of interest.

2.1 Attribute-Related Information

Mandatory information to be specified for each attribute is: name, data type (string, long text, integer, float) and dimensionality (e.g. 2 for screen resolution like 800x600). In order to automatically extract an attribute, additional knowledge is typically needed. Extraction knowledge about the attribute *content* includes (1) textual value patterns; (2) for integer and float types: min/max values, a numeric value distribution and possibly units of measure; (3) value length in tokens: min/max length constraints or a length distribution; (4) axioms expressing more complex constraints on the value and (5) coreference resolution knowledge. Attribute *context* knowledge includes (1) textual context patterns and (2) formatting constraints.

Patterns in *Ex* (for both the value and the context of an attribute or class) are nested regular patterns defined at the level of tokens (words), characters, formatting tags (HTML) and labels provided by external tools. Patterns may be inlined in the extraction ontology or sourced from (possibly large) external files, and may include the following:

- specific tokens, e.g. 'employed by'
- token *wildcards*, which require one or more token properties to have certain values (e.g. any capital or uppercase token, any token whose lemma is 'employ')
- *character-level* regular expressions for individual tokens
- *formatting tags* or their classes, such as 'any HTML block element'
- *labels* that represent the output of external tools, such as sentence boundaries, part-of-speech tags, parsed chunks or output from other IE engines.
- *references* to other *matched patterns*; this allows for construction of complex grammars where rules can be structured and reused
- *references* to other matched *attribute candidates*: a *value* pattern containing a reference to another attribute means that it can be nested inside this attribute's value; for *context* patterns, attribute references help encode how attributes follow each other

For *numeric* types, default value patterns for integer/float numbers are provided. Linking a numeric attribute to unit definitions (e.g. to currency units) will automatically create value patterns containing the numeric value surrounded by the units.

For both attribute and class definitions, *axioms* can be specified that impose constraints on attribute value(s). Axioms defined for a single attribute are used to boost or suppress confidence scores of candidate attribute values based on whether the axiom is satisfied or not for those values. For a class, each axiom may refer to all attribute values present in the partially or fully parsed instance. For example, a price with tax must be greater than the price without tax. Axioms can be authored using the JavaScript[3] scripting language. We chose JavaScript since it allows arbitrarily complex axioms to be constructed and because the web community is used to it.

In addition, *formatting constraints* may be provided for each attribute. Currently, four types of formatting constraints are supported: (1) the whole attribute value is contained in a single parent, i.e. it does not include other tags or their boundaries; (2) the value fits into the parent; (3) the value does not cross any inline formatting elements; (4) it does not cross any block elements. We investigate how custom constraints could easily be added by users. By default, all four constraints are in effect and influence the likelihood of attribute candidates being extracted.

In many tasks, it is necessary to identify co-referring occurrences of attribute values. For each attribute, a coreference resolution script can be provided that determines whether two values of the same attribute (or of its extensions) may co-refer to the same entity. By default, identical extracted values for the same attribute are treated as co-referring.

2.2 Class-Related Information

Each *class definition* enumerates the attributes which may belong to it, and for each attribute it defines a *cardinality* range and optionally a cardinality distribution. Extraction knowledge may address both content and context of the class. *Class content patterns* are analogous to the attribute value patterns, however, they may match *parts* of an instance and must contain at least one *reference* to a member attribute. Class content patterns may be used e.g. to describe common wordings used between attributes or just to specify attribute ordering. *Class context patterns* are analogous to attribute context patterns.

Axioms are used to constrain or boost instances based on whether their attributes satisfy the axiom. For each attribute, an *engagedness* parameter may be specified to estimate the apriori probability of the attribute joining a class instance (as opposed to standalone occurrence). Regarding class context, analogous *class context patterns* and similar *formatting constraints* as for attributes are in effect also for classes.

In addition, constraints can be specified that hold over the whole sequence of extracted objects. Currently supported are minimal and maximal instance counts to be extracted from a document for each class.

2.3 Extraction Evidence Parameters

All types of extraction knowledge mentioned above, i.e. value and context patterns, axioms, formatting constraints and ranges or distributions for numeric attribute values and for attribute content lengths, are essentially pieces of evidence indicating the presence (or absence) of a certain attribute or class instance. In *Ex*, every piece of evidence may be equipped with two probability estimates: precision and recall. The *precision* of evidence states how probable it is for the predicted attribute or class

---

[3] http://www.mozilla.org/rhino

instance to occur given the evidence holds, disregarding the truth values of other evidence. For example, the precision of a left context pattern "person name: $" (where $ denotes the predicted attribute value) may be estimated as 0.8; i.e. in 80% of cases we expect a person name to follow in text after a match of the "person name:" string. The *recall* of evidence states how abundant the evidence is among the predicted objects, disregarding whether other evidence holds. For example, the "person name: $" pattern could have a low recall since there are many other contexts in which a person name could occur.

Pattern precision and recall can be estimated in two ways. First, annotated documents can be used to estimate both parameters using simple ratios of counts observed in text. In this case, it is necessary to smooth the parameters using an appropriate method. For a number of domains it is possible to find existing annotated data, e.g. web portals often make available online catalogs of manually populated product descriptions linking to the original sellers' web pages. When no training data is available or if the evidence seems easy to estimate, the user can specify both parameters manually. For the experimental results reported below we estimated parameters manually.

2.4 Extraction Ontology Samples

In order to illustrate most of the above features, we present and explain two relatively large real-world examples from two different domains.[4]

The first example is from the *product catalogue* domain. Fig. 1 shows the structure of an extraction ontology for the domain of computer monitor offers. Fig. 2 displays part of the corresponding code in the XML-based ontology definition language, dealing with the name of the monitor model, its price and derived prices with and without tax.

In the global scope of the model, extraction knowledge affecting more than one attribute is defined. First, a pattern declares that in 75% of cases a *Monitor* instance starts with its *name* which is closely (within at most 20 tokens) followed by up to 4 *price* attributes. Second, an axiom says that in all cases the *price_with_tax* must be greater than *price_without_tax* (if both are specified).

The textual *name* attribute shows the usage of generic patterns (model_id) which can be reused within the following value and context sections. The first pattern in the value section describes a typical form of a computer monitor name and claims that 80% of its matches identify a positive example of monitor name. It also claims that 75% of monitor names exhibit this pattern. The value section further constrains the name's length and its position within HTML formatting elements; e.g. the last formatting pattern denies crossing of formatting block boundaries by saying that 100% of monitor names do not cross block tags. The context section's only pattern says that in 20% of cases, monitor names are preceded by labels like 'model name' and that observing such labels identifies monitor names with a 30% precision only.

---

[4] More details about experiments carried out in those domains are in Section 4.

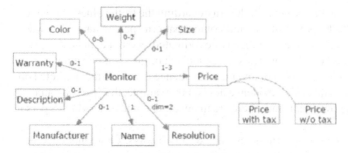

Fig. 1 General scheme of extraction ontology of monitors

```
<class id="Monitor">

<pattern id="name_first_and_price_follows" type="pattern" cover="0.75">
  $name (<tok/>{0.20} $price){1,4}  </pattern>

<axiom> $price_with_tax > $price_without_tax </axiom>

<attribute id="name" type="name" card="1" eng="0.70">
<pattern id="model_id">
  <tok case="UC"/> | <tok type="ALPHA" case="CA"/> | <tok type="ALPHANUM|INT"/>
</pattern>
<value>
  <pattern cover="0.5" p="0.8">  (LCD (monitor|panel)?)?
    <pattern src="manuf.txt"/> <pattern ref="model_id"/>{1,2}
  </pattern>
  <length><distribution min="1" max="7"/></length>
  <pattern cover="0.5" type="format"> has_one_parent </pattern>
  <pattern cover="0.5" type="format"> fits_in_parent </pattern>
  <pattern cover="0.8" type="format"> no_crossed_inline_tags </pattern>
  <pattern cover="1.0" type="format"> no_crossed_block_tags </pattern>
</value>
<context>
  <pattern p="0.3" cover="0.2"> model? name :? $ </pattern> </context>
</attribute>

<attribute id="price" type="float" card="1-4" units="euro, pound, dollar" eng="0.80">
<value>
  <pattern cover="0.9" p="0.9">
    <pattern ref="generic.price"/> $unit | $unit <pattern ref="generic.price"/> </pattern>
    <distribution min="100" max="5000"/>
    <transform> $.replace(/\s+/,"") </transform>
  </pattern>
<context>
  <pattern cover="0.05" p="0.5"> price of? :? $ </pattern> </context>
</attribute>

<attribute id="price_with_tax" card="0-1" type="float" extends="price">
<pattern id="intro">
price? ( (with|including|incl.) (tax|taxes) | \(? (tax|taxes) included \)? )
</pattern>
<pattern id="outro">
( (with|including|incl.) (tax|taxes) | \(? (tax|taxes) included \)? | \( price with (tax|taxes) \) )
</pattern>
<context>
  <pattern cover="0.5" p="0.8"> $ <pattern ref="outro"/> </pattern>
  <pattern cover="0.1" p="0.1"> <pattern ref="intro"/> :? $ </pattern>
</context>
</attribute>
```

Fig. 2 Fragment of code of extraction ontology for computer monitors

The *price* attribute and its two extensions are numeric. The first pattern of the value section utilises a pattern 'generic.price' and currency units, both of which are imported from a generic datatypes model (not shown). The numeric value constraint uses by default the first unit (Euro). The value transformation is a script applied to the value after extraction. The *price_with_tax* attribute, and similarly,

```
<class id="Contact">

<script src="contact.js" />

<pattern id="title_name_together_and_first" type="pattern" cover="0.7">
 ^ $title{0-3} .? $name .? $title{0-3} </pattern>

<axiom cover="0.8"> nameMatchesEmail($name, $email) > 0 </axiom>

<classifier id="cls1" method="weka" classtype="attribute"
 name="weka.classifiers.rules.JRip" features="contact_all_prip.feat"
 model="../ex/data/med/train/contact_all_jrip.bin" elements="*"/>

<attribute id="degree" type="name" card="0-4" eng="0.60">
 <pattern id="degree_pre"> ( Miss | Lady | Sir ) .? </pattern>
 <pattern id="degree_suf"> ( MSc | MA | MPh ) .? </pattern>
 <value>
  <pattern cover="0.7" p="0.8" ignore="case">
   <pattern ref="degree_pre" /> | <pattern ref="degree_suf"/> </pattern>
  <pattern cover="1" type="format"> has_one_parent </pattern>
 </value>
</attribute>

<attribute id="name" type="name" card="1" eng="0.80">
 <pattern id="init"> (A|B|C|D|E|F|G|H|I|J|K|L|M|N|O|P|R|S|T|U|V|W|X|Y|Z) .? </pattern>
 <value>
  <pattern cover="0.5" p="0.5">
   <pattern src="first.txt" type="pattern" ignore="lemma" case="CASIC" />
   <pattern ref="init" />?
   <pattern src="last.txt" type="pattern" ignore="lemma" case="CASIC" />
  </pattern>
  <pattern cover="0.25" p="0.4">
   <pattern src="first.txt" type="pattern" ignore="lemma" case="CASIC" />
   <pattern ref="init" />?
   <tok type="alpha" case="CASIC"/>
  </pattern>
  <pattern p="0.7" cover="0.5"> <phr label="cls1.name" /> </pattern>
  <pattern cover="0.3" type="format"> fits_in_parent </pattern>
  <length> <distribution min="1" max="6" /> </length>
  <axiom> checkPersonName($) </axiom>
  <refers> nameRefersTo($, $other) </refers>
 </value>
 <context> <pattern cover="0.05" p="0.6"> person? name :? $ </pattern> </context>
</attribute>

<attribute id="email" type="name" card="0-1" eng="0.60">
 <value> <pattern cover="0.90" p="0.9"> <pattern ref="generic_email"/> </pattern>
 </value>
</attribute>
</attribute>
```

Fig. 3 Fragment of code of extraction ontology for contact information

price_without_tax (not shown) inherit all extraction knowledge from *price* and specify additional context patterns. Observing these patterns will cause the price being extracted as one of the two extensions.

The second example is taken from the *contact information* ontology developed within the EU project MedIEQ;[5] the goal of the project is to ease expert-based accreditation of medical websites by automatically extracting information that is critical for the evaluation. This includes, among other, *contact information* of website responsible. A simplified version of the code related to persons' degrees and emails is shown.

In the global scope of the model, we can see the extraction knowledge referring to more than one attribute. The first pattern states that in 70% of cases, a *Contact* starts with its name or degree, and that these are only separated by punctuation. The axiom claims that in 80% of cases, the person's name and email exhibit some string similarity which can be intercepted by the referenced *script function* nameMatchesEmail() which returns non-zero if it thinks the given name corresponds to the

---

[5] http://www.medieq.org

given email. This function has been imported from a *script* "contact.js" above. Finally, a *classifier link* contracts an external trained classifier to classify all attributes of the Contact class. Classifications made by this classifier can be used in all patterns of this ontology.

The *degree* attribute demonstrates how generic patterns (degree_pre and degree_suf) can be reused by other patterns. The value section declares that in 80% of cases, observing the specified pattern really identifies a degree. It also claims that 70% of all degree occurrences will match this pattern (much larger enumeration would be needed to satisfy this number in reality). The *formatting pattern* requires all degrees to be enclosed in a single formatting element.

The *name* attribute's value section draws extraction knowledge from several sources. Its first pattern uses large first name and surname lists and inserts an optional initial in between. It claims to be relatively precise: 80% of its matches should correctly identify a person name. However, it is only expected to cover about 40% of all person names as the lists can never be exhaustive. The second pattern is less precise as it allows any alphabetic surname which is either capital or uppercase. When this pattern is matched, the extraction decision will strongly depend on the observed values of other extraction evidence. The third pattern takes into account the classifier's positive decision and it expects the classifier to have a 70% precision and a 50% recall. In addition, the name length is constrained and an axiom can perform a final check of the to-be-extracted value using a script function. As there are often multiple occurrences of a single person name on a web page, the refers section uses a script function to determine whether two names may co-refer (e.g. John Smith to Smith). The context section describes the typical surrounding phrases. As the context pattern's precision is not enough to create new person names, it will boost person name candidates for which it is observed (thanks to its precision) and suppress a little those for which it is missing (thanks to its coverage).

For the *email* attribute, just a single pattern is used within its value section which refers to a generic pattern defined by an imported datatype extraction ontology.

3   The Extraction Process

The inputs to the extraction process are the extraction ontology and a set of documents. Extraction consists of six stages depicted in Fig. 4.

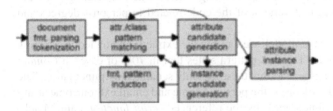

Fig. 4 Extraction process schema

3.1 Document Preprocessing

Initially, the text of each analysed document is tokenised (including assignment of different token types) and, for HTML documents, their formatting structure is converted to a tree of formatting labels (using the CyberNeko HTML parser[6]) that contain the tokenised text. Optionally, lemmatisation, sentence boundary classification, part-of-speech tagging and further third-party tools can be run in this phase to create labels over the tokenised document that can be used in later stages.

3.2 Attribute Candidate Generation

After loading a document, all attribute value and attribute context patterns of the ontology are matched against the document's tokens. Where a value pattern matches, the system attempts to create a new candidate for the associated attribute (attribute candidate – AC). If more value patterns match at the same place, or if there are context pattern matches for this attribute in neighbouring areas, then the corresponding evidence is turned on as well for the AC. Evidence corresponding to all other non-matched patterns is kept off for the AC. Also, during the creation of the AC, all other evidence types (axioms, formatting constraints, content length and numeric value ranges) are evaluated and set. The set of all evidence Φ_A known for the attribute A is used to compute a *conditional probability estimate* P_{AC} of how likely the AC is given all the observed evidence values:

$$P_{AC} = P(A|E \in \Phi_A) \tag{1}$$

The full formula is described and derived in [6]. We assume conditional independence of evidence given that the attribute holds or does not hold. The AC is created only if P_{AC} exceeds a pruning threshold defined by the extraction ontology.

In places where a context pattern matches and there are no value pattern matches in neighbourhood, the system tries to create ACs of various length (in tokens) in the area pointed to by the context pattern. Again, all the evidence values are combined to compute P_{AC} for each new AC. For patterns which include other attributes, we run the above process until no new ACs are generated.

The set of (possibly overlapping) ACs created during this phase is represented as an *AC lattice* going through the document, where each AC is scored by $score(AC) = log(P_{AC})$. Apart from the ACs which may span multiple tokens, the lattice also includes one 'background' state for each token that takes part in some AC. A background state BG_w for token w is scored as follows:

$$score(BG_w) = \min_{AC, w \in AC} log(\frac{1 - P(AC)}{|AC|}) \tag{2}$$

---

[6] http://people.apache.org/~andyc/neko/doc/html/

where $|AC|$ is the length of the AC in tokens. The extraction process can terminate here if no instance generation or formatting pattern induction is done, in which case all ACs on the best path through the lattice are extracted.

3.3 Instance Candidate Generation

At the beginning of the instance candidate (IC) generation phase, each AC is used to create a simple IC consisting just of that single AC. Then, a bottom-up IC generation algorithm is employed to generate increasingly complex ICs from the working set of ICs. At each step, the highest scoring (seed) IC is chosen and its neighbourhood is searched for ACs that could be added to it without breaking ontological constraints for the IC class. Only a *subset* of the constraints is taken into account at this time as e.g. some minimum cardinality constraints or axioms requiring the presence of multiple attributes could never get satisfied initially. Each added AC is also examined to see whether it may co-refer with some AC that is already present in the IC; if yes, it is only added as a reference and it does not affect the resulting IC score.

After adding ACs to the chosen seed IC, that IC is removed from the working set and the newly created larger ICs are added to it. The seed IC is added to a *valid IC set* if it satisfies *all* ontological constraints. As more complex ICs are created by combining simpler ICs with surrounding ACs, a limited number of ACs or AC fragments is allowed to be skipped (AC_{skip}) between the combined components, leading to a penalization of the created IC. The IC scores are computed based on their AC content and based on the observed values of evidence E known for the IC class C:

$$sc_1(IC) = exp\left(\frac{\sum_{AC \in IC} log(P_{AC}) + \sum_{AC_{skip} \in IC}(1 - log(P_{AC_{skip}}))}{|IC|}\right) \qquad (3)$$

$$sc_2(IC) = P(C|E \in \Omega_C) \qquad (4)$$

where $|IC|$ is the number of member ACs and Ω_C is the set of evidence known for class C; the conditional probability is estimated as in Eq. 1. By experiment we chose the Prospector [2] pseudo-bayesian method to combine the above into the final IC score:

$$score(IC) = \frac{sc_1(IC)sc_2(IC)}{sc_1(IC)sc_2(IC) + (1 - sc_1(IC))(1 - sc_2(IC))} \qquad (5)$$

The IC generation algorithm picks the best IC to expand using the highest $score(IC)$. The generation phase ends when the working set of ICs becomes empty or on some terminating condition such as after a certain number of iterations or after a time limit has elapsed. The output of this phase is the set of valid ICs.

As the generated IC space can get very large for some extraction ontologies, it can be constrained by several configurable pruning parameters: the IC probability can be thresholded so that only promising hypotheses are generated; the absolute beam size limits the number of most probable ICs to be kept for each span in the document; the relative beam size is used to prune all ICs within a span whose probability relative to the best IC in that span is less than the specified ratio.

3.4 Formatting Pattern Induction

During the IC generation process, it may happen that a significant part of the created valid ICs satisfies some (apriori unknown) *formatting pattern*. For example, a contact page may consist of 6 paragraphs where each paragraph starts with a bold person name and scientific degrees. A more obvious example would be a table with the first two columns listing staff first names and surnames. Then, if e.g. 90 person names are identified in such table columns and the table has 100 rows, the induced patterns make the remaining 10 entries more likely to get extracted as well.

Based on the lattice of valid ICs, the following *pattern induction* procedure is performed. First, the best scoring sequence of non-overlapping ICs is found through the lattice. Only the ICs on the best path take part in pattern induction. For each IC, we find its nearest containing formatting block element. We then create a subtree of formatting elements between the containing block element (inclusive) and the attributes comprising the IC. This subtree contains the names of the formatting elements (e.g. paragraph or bold text) and their order within parent (e.g. the first or second cell in table row). Relative frequencies of these subtrees are calculated over the examined IC set (separately for each class if there are more). If the relative and absolute frequencies of a certain subtree exceed respective configurable thresholds, that subtree is converted to a new, *local* context pattern that indicates the presence of the corresponding class. Such locally induced evidence is only used within the currently analysed document or within a group of documents that share common formatting. The precision and recall of the induced context patterns are based on the relative frequencies with which the patterns hold in the document (or document set) with respect to the observed ICs.

The newly created context patterns are then fed back to the pattern matching phase, where they are matched and applied. This extra iteration rescores existing ACs and ICs and may as well yield new ACs and ICs which would not have been created otherwise. With our current implementation we have so far only experimented with pattern induction for ICs composed of a single attribute. Using this feature typically increases recall but may have adverse impact on precision. One approach to avoid degradation of precision is to provide attribute evidence which will prevent unwanted attributes from being extracted.

3.5 Attribute and Instance Parsing

The purpose of this final phase is to output the most probable sequence of instances and standalone attributes through the analysed document. The valid ICs are merged into the AC lattice described in Section 3.2 so that each IC can be avoided by taking a competing path through standalone ACs or through background states. In the merged lattice, each IC is scored by $log(score(IC))|IC|$.

The merged lattice is searched using *dynamic programming* for the most probable sequence of non-overlapping extractable objects. Path scores are computing by adding the state scores as they consist of log probability estimates. To support *n-best* results, up to *n* back pointers, ordered by their accumulated scores, are stored for each visited state. The *n* best paths are retrieved by backtracking using an A^*-based algorithm [4]. Extractable objects (attribute values and instances) can be read from each extracted path and sent to output.

In order to support constraints over the whole extracted sequence of objects (such as the minimal and maximal instance counts in a document), the search algorithm has been extended so that back pointers accumulated for each visited state include information about how the particular partial path leading into the visited state (inclusive) satisfies the constraint. In case of the instance count constraint, each back pointer of each visited state includes the instance counts observed on the corresponding incoming partial path. Paths that violate the constraints in a way that cannot be undone by further expansion of the path are not prolonged by the search. All paths that reach the final state of the document lattice and do not completely satisfy the constraints are discarded.

4   Experimental Results

4.1   *Contact Information on Medical Pages*

In the EU (DG SANCO) MedIEQ project[7] we experiment with several dozens of *medical website quality criteria*, most of which are to be evaluated with the assistance of IE tools. One of them is the presence and richness of contact information. Results are presented for 3 languages. A meta-search engine [14] was used to look up relevant medical web sites; these websites were then spidered, their contact pages were manually identified, contact information was annotated and the pages were added to contact page collections for each language. In total, 109 HTML pages were assembled for English with 7000 named entities, 200 for Spanish with 5000 and 108 for Czech with 11000 named entities. A contact extraction ontology was developed for each language, with language-independent parts reused. The extraction ontology developer was allowed to see 30 randomly chosen documents from each collection for reference and to use any available gazetteers such as city names, frequent first names and surnames. Results were evaluated using the remaining documents. On average, each ontology contained about 100 textual patterns for the context and content of attributes and of the single extracted 'contact' class, attribute length distributions, several axioms and co-reference resolution rules. The effort spent on developing and tuning the initial English ontology was about 2-3 person-weeks, and its customization to a new language amounted to 2 person-weeks.

---

[7] http://www.medieq.org

Table 1 Contact IE results for three languages

	English			Spanish			Czech		
attribute	prec	recall	F	prec	recall	F	prec	recall	F
title	71/78	82/86	76/82	-	-	-	87/89	88/91	88/90
name	66/74	51/56	58/64	71/77	81/86	76/81	74/76	82/83	78/80
street	62/85	52/67	56/75	71/93	46/58	56/71	78/83	66/69	71/75
city	47/48	73/76	57/59	48/50	77/80	59/61	67/75	69/79	68/77
zip	59/67	78/85	67/75	88/91	91/94	89/93	91/91	97/97	94/94
country	58/59	89/89	70/71	67/67	78/78	72/72	64/66	87/96	74/78
phone	97/99	84/87	90/93	84/89	91/96	87/92	92/93	85/85	88/89
email	100/100	99/99	100/100	94/95	99/99	96/97	99/99	98/98	98/98
company	57/81	37/51	44/63	-	-	-	-	-	-
dept.	51/85	31/45	38/59	-	-	-	-	-	-
overall	70/78	62/68	66/72	71/76	81/86	76/80	81/84	84/87	82/84

Fig. 5 Sample of automatically annotated data; extracted instances on the right

Table 1 shows contact extraction results for 3 languages.[8] Two evaluation modes are used: strict (the first number) and loose. In the strict mode of evaluation, only exact matches are considered to be successfully extracted. In the loose mode, partial credit is also given to incomplete or overflown matches; e.g. extracting 'John Newman' where 'John Newman Jr.' was supposed to be extracted will count as a 66% match (based on overlapping word counts). Some of the performance numbers below may be impacted by relatively low inter-annotator agreement. Fig. 5 shows a sample of automatically annotated data.

4.2 Product Catalogues

Another application of the *Ex* system is to extract information about products sold or described online. Our experiments have so far been limited to TV sets, computer monitors and bicycles. For each product type, an initial version of extraction

---

[8] At the time of writing, degrees were not annotated as part of the Spanish collection and results for company and department names for Spanish and Czech were still work in progress.

ontology has been created to extract instances composed of typically a model name, price and multiple product-specific attributes. For example, the monitor extraction ontology contains 11 attributes, about 50 patterns and several axioms. The effort spent so far is about 2 person weeks. We experiment with a data set of 3,000 partially annotated web pages containing monitor ads. The average F-measure is now around 75% but complete evaluation has not been completed yet.

In order to obtain annotated data, we cooperate with one of the largest Czech web portals. The annotated data come from the original websites some of which send structured data feeds to the portal in order to get included in their online product database. The portal, on the other hand, can use such data to develop and train extraction models to cover the remaining majority of sites.

4.3 Weather Forecasts

Finally, we experimented with the domain of weather forecasts. Here our goal was to investigate the possibility to assist the ontology engineer in reusing existing *domain ontologies* in order to develop the extraction one/s. An advantage of this domain was the fact that several OWL ontologies were available for it. We analysed three of them by means of applying generic rules of two kinds:

1. Rules suggesting the *core class/es* for the extraction ontology. As the extraction ontology for extraction from HTML-formatted text[9] is typically more class-centric and hierarchical than a properly-designed domain ontology, only few classes from the domain ontology are likely to become classes in the extraction ontology, while others become attributes that are dependent on the core class/es. For example, 'Day' is typically an attribute of a 'Forecast' class in an extraction ontology, while in the domain ontology they could easily be two classes connected by a relationship. One of such core class selection rules is, in verbal form, e.g. "Classes that appear more often in the domain than in the range of object properties are candidates for core class/es.".
2. Rules performing the actual *transformation*. Examples of such rules are e.g. "A data type property D of class C may directly yield an attribute of C." or "A set of mutually disjoint subclasses of class C may yield an attribute, whose values are these subclasses."

Most such independently formulated selection and transformation rules appeared as performing well in the initial experiment in the weather domain; details are in [10]. Transformation rules seemed, by first judgement, to suggest a sensible and inspiring, though by far not complete, skeleton of an extraction ontology. Testing this ontology on real weather forecast records is however needed for proper assessment.

In general, although the first experiments look promising, extensive usage of domain ontologies as starting point for extraction ontologies seems to be hindered by unavailability of high-quality domain ontologies for most domains, e.g. in relation to different categories of products or services, judging by the results of

---

[9] This is not the case for extraction from free text, which is more relation-centric.

Swoogle-based[10] retrieval. This obstacle is likely to disappear in the not-so-distant future, as the semantic web technology becomes more widespread.

4.4 Discussion

The main practical advantage of the approach based on extraction ontologies was expected to be the *rapid start* of the whole process while, at the same time, the growing body of extraction knowledge remains manageable in long term thanks to support for high-level *conceptual modelling*. This assumption seems to have been verified by our experience from the mentioned applications.

Preliminary performance comparisons with other IE systems showed that our approach was slightly below the top-scoring systems for a common benchmark task of *seminar announcement* extraction. Unlike most purely inductively trained approaches, extraction ontologies have the advantage of being applicable to domains with limited or no training data at hand. Initial evaluation results of purely manual extraction ontologies and extraction ontologies coupled with trainable classifiers are reported at [6] for several tasks.

5 Related Work

Most state-of-the-art WIE approaches focus on identifying structured collections of items (records), typically using inductively learnt models. Ontologies are often considered but rather as additional structures to which the extracted data are to be adapted after they have been acquired from the source documents, for the sake of a follow-up application [5]. There is no provision for directly using the rich structure of a domain-specific ontology in order to guide the extraction process.

Though our system has an optional wrapper-like feature, it also significantly differs from mainstream *wrapper* tools such as Kapow[11] or Lixto[12], which focus on building wrappers for each page/site separately, while our extraction ontologies can be reused within the whole domain. Among the wrapper-oriented approaches, the most similar to ours seems to be HiLεX [12], which allows to specify extraction ontologies as trees of linguistic and structural elements and evaluate them using a powerful logical language. Due to its assumption of nested rectangular semantic portions of web pages, it is however tuned to extraction from tabular data (rather than from more linearly-structured data), although it relaxes the dependence on HTML formatting proper.

The approach to WIE that is inherently similar to ours (and from which we actually got inspiration in the early phase of our research) is that developed by Embley and colleagues at BYU [3]. The main distinctive features of our approach are: (1) the possibility to provide the extraction patterns with probability estimates (plus

---

[10] http://swoogle.umbc.edu

[11] http://www.kapowtech.com

[12] http://www.lixto.com

other quantitative info such as value distributions), allowing to calculate the weight
for every attribute candidate as well as instance candidate; (2) the effort to combine hand-crafted extraction ontologies with other sources of information—HTML
formatting and/or known data instances (3) the pragmatic distinction between extraction ontologies and domain ontologies proper: extraction ontologies can be arbitrarily adapted to the way domain data are typically *presented* on the web while
domain ontologies address the domain as it is (but can be used as starting point for
designing extraction ontologies). For similarly pragmatic reasons (easy authoring),
we also used a proprietary XML syntax for extraction ontologies. An objective comparison between both approaches would require detailed experiments on a shared
reference collection.

An approach to automatically discover new extractable attributes from large
amounts of documents using statistical and NLP methods is described in [11].
On the other hand, formatting information is heavily exploited for IE from tables
in [16]. Our system has a slightly different target; it should allow for fast IE pro-
totyping even in domains where there are few documents available and the content
is semi-structured. While our system relies on the author to supply coreference resolution knowledge for attribute values, advanced automatic methods are described
e.g. in [18]. The system described in [17] uses statistical methods to estimate the
mutual affinity of attribute values.

Our ideas and experiments on domain ontology selection and transformation to
extraction ontology are related to the generic research in ontology selection and content evaluation [13], especially with respect to the notion of intra-ontology concept
centrality; this relationship deserves further study.

6   Conclusions

The *Ex* system attempts to unify the often separate phases of WIE and ontology
population. Multiple sources of extraction knowledge can be combined: manually encoded knowledge, knowledge acquired from annotated data and knowledge
induced from common formatting patterns by the means of wrapper induction. An
alpha version of *Ex* (incl. extraction ontology samples) is publicly available[13].

Future work will evaluate the system integrated with *trainable* machine learning
algorithms and exploit some more coarse-grained web mining tools including *web
page classifiers*. For the latter, a rule-based *post-processing engine* is under development; it will, following a higher-level domain-specific website model, filter and
transform the extraction results based on the context—semantic class of the given
page. Both *instance parsing* and *formatting pattern induction* algorithms themselves
also need improvement in accuracy and speed. Furthermore, we plan to investigate
how *text mining* over the extraction results could help us identify 'gaps' in the ontology, e.g. non-labelled tokens frequently appearing inside a 'cloud' of annotations

---

[13] http://eso.vse.cz/~labsky/ex

are likely to be unrecognised important values. Finally, we intend to provide support for semi-automated transformation of *domain ontologies* to extraction ones.

Acknowledgements. The research leading to this paper was partially supported by the EC under contract FP6-027026, Knowledge Space of Semantic Inference for Automatic Annotation and Retrieval of Multimedia Content - K-Space. The medical website application is carried out in the context of the EC-funded (DG-SANCO) project MedIEQ.

References

1. Ciravegna, F.: (LP)2, an Adaptive Algorithm for Information Extraction from Web-related Texts. In: Proc. IJCAI 2001 Workshop on Adaptive Text Extraction and Mining, Seattle (2001)
2. Duda, R.O., Gasching, J., Hart, P.E.: Model design in the Prospector consultant system for mineral exploration. In: Readings in Artificial Intelligence, pp. 334–348 (1981)
3. Embley, D.W., Tao, C., Liddle, D.W.: Automatically extracting ontologically specified data from HTML tables of unknown structure. In: Spaccapietra, S., March, S.T., Kambayashi, Y. (eds.) ER 2002. LNCS, vol. 2503, pp. 322–337. Springer, Heidelberg (2002)
4. Huang, X., Acero, A., Hon, H.W.: Spoken Language Processing: A Guide to Theory, Algorithm and System Development. Prentice Hall, New Jersey (2001)
5. Kiryakov, A., Popov, B., Terziev, I., Manov, D., Ognyanoff, D.: Semantic annotation, indexing, and retrieval. J. Web. Sem. 2, 49–79 (2004)
6. Labský, M.: Information Extraction from Websites using Extraction Ontologies. Technical Report, KEG UEP (2009), http://eso.vse.cz/~labsky/ex/exo09.pdf
7. Labský, M., Svátek, V.: On the Design and Exploitation of Presentation Ontologies for Information Extraction. In: ESWC 2006 Workshop on Mastering the Gap: From Information Extraction to Semantic Representation. CEUR-WS, vol. 187 (2006)
8. Labský, M., Svátek, V., Šváb, O.: Types and Roles of Ontologies in Web Information Extraction. In: ECML/PKDD Workshop on Knowledge Discovery and Ontologies, Pisa (2004)
9. Lafferty, J., McCallum, A., Pereira, F.: Conditional random fields: Probabilistic models for segmenting and labeling sequence data. In: Proc. 18th International Conf. on Machine Learning, pp. 282–289. Morgan Kaufmann, San Francisco (2001)
10. Nekvasil, M., Svátek, V., Labský, M.: Transforming Existing Knowledge Models to Information Extraction Ontologies. In: Proc. 11th International Conference on Business Information Systems. LNBIP, vol. 7, pp. 106–117. Springer, Heidelberg (2008)
11. Popescu, A., Etzioni, O.: Extracting Product Features and Opinions from Reviews. In: Proc. Conference on Human Language Technology and Empirical Methods in Natural Language Processing, Vancouver, Canada, pp. 339–346 (2005)
12. Ruffolo, M., Manna, M.: HiLεX: A System for Semantic Information Extraction from Web Documents. In: Proc. Enterprise Information Systems. LNBIP, pp. 194–209. Springer, Heidelberg (2008)
13. Sabou, M., Lopez, V., Motta, E.: Ontology selection for the real semantic web: How to cover the queen's birthday dinner? In: Staab, S., Svátek, V. (eds.) EKAW 2006. LNCS, vol. 4248, pp. 96–111. Springer, Heidelberg (2006)

14. Stamatakis, K., Metsis, V., Karkaletsis, V., Růžička, M., Svátek, V., Amigó, E., Pöllä, M., Spyropoulos, C.D.: Content collection for the labelling of health-related web content. In: Bellazzi, R., Abu-Hanna, A., Hunter, J. (eds.) AIME 2007. LNCS, vol. 4594, pp. 341–345. Springer, Heidelberg (2007)

15. Svátek, V., Labský, M., Vacura, M.: Knowledge Modelling for Deductive Web Mining. In: Motta, E., Shadbolt, N.R., Stutt, A., Gibbins, N. (eds.) EKAW 2004. LNCS, vol. 3257, pp. 337–353. Springer, Heidelberg (2004)

16. Wei, X., Croft, B., McCallum, A.: Table Extraction for Answer Retrieval. Information Retrieval Journal 9(5), 589–611 (2006)

17. Wick, M., Culotta, A., McCallum, A.: Learning Field Compatibilities to Extract Database Records from Unstructured Text. In: Proc. Conference on Empirical Methods in Natural Language Processing, Sydney, Australia, pp. 603–611 (2006)

18. Yates, A., Etzioni, O.: Unsupervised Resolution of Objects and Relations on the Web. In: Proc. NAACL Human Language Technologies Conference, pp. 121–130 (2007)

Dealing with Background Knowledge in the SEWEBAR Project

Jan Rauch and Milan Šimůnek

Abstract. SEWEBAR is a research project the goal of which is to study possibilities of dissemination of analytical reports through Semantic Web. We are interested in analytical reports presenting results of data mining. Each analytical report gives answer to one analytical question. Lot of interesting analytical questions can be answered by GUHA procedures implemented in the LISp-Miner system. The SEWEBAR project deals with these analytical questions. However the process of formulating and answering such analytical questions requires various background knowledge. The paper presents first steps in storing and application of several forms of background knowledge in the SEWEBAR project. Examples concerning dealing with medical knowledge are presented.

1 Introduction

SEWEBAR (SEmantic WEB and Analytical Reports) [16, 21] is an academic research project developed at Faculty of Informatics and Statistics of University of Economics, Prague. The project is inspired by *10 challenging problems in data mining research* see http://www.cs.uvm.edu/~icdm/. One of them is characterized as *mining complex knowledge from complex data*. Necessity to relate data mining results to the real world decisions they affect is emphasized in relation to this problem. One way how to meet this requirement is to arrange results of data mining into an analytical report structured both according to an analyzed problem and to user's needs. The background knowledge must be taken into consideration when

Jan Rauch and Milan Šimůnek
Faculty of Informatics and Statistics, University of Economics, Prague
e-mail: {rauch,simunek}@vse.cz

B. Berendt et al. (Eds.): Knowl. Disc. Enhan. with Sem. and Soc. Info., SCI 220, pp. 89–106.
springerlink.com © Springer-Verlag Berlin Heidelberg 2009

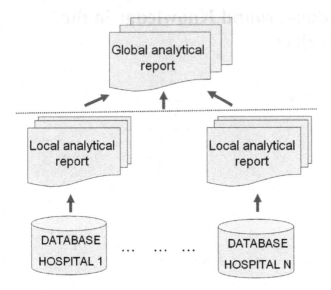

Fig. 1 Sharing analytical reports on Semantic Web

preparing the report. An attempt to produce such analytical report automatically is described in [11], these possibilities are also discussed in [14]. Such analytical reports are natural candidates for Semantic Web [8]. A possible result of development in this direction is outlined in Fig. 1 [16, 21].

There are hospitals storing data concerning patients in their databases. Automatically (or semi-automatically) produced *local analytical reports* give answers to *local analytical questions* concerning patients, their illnesses and treatments in particular hospitals. There are also (semi)automatically produced *global analytical reports*. Each global analytical report summarizes two or more local analytical reports. It is supposed that content of analytical reports is indexed among other by logical formulas corresponding to patterns resulting from data mining instead of usual key words. The possibility of indexing content of analytical reports by logical formulas is outlined in [14], see also [7, 9].

Main features of the SEWEBAR project can be summarized as follows [16, 21].

- SEWEBAR is the research project. Its goal is to study possibilities of creation and dissemination of analytical reports presenting results of data mining through Semantic Web. The whole project is in an early stage. It is however based on long time development of the LISp-Miner academic software system [18, 19, 28] that is used both in applications and in teaching of data mining [24, 23].
- The core of the LISp-Miner system consists of six GUHA procedures [19, 20, 24]. Input of each GUHA procedure is given by a definition of a set of relevant patterns and by analyzed data. Output of the procedure is a set of all prime patterns. The pattern is prime if it is relevant and true in the given data and if it does not logically follow from an other more simple output pattern.

- The usual apriori algorithm is not used in the GUHA procedures of the LISp-Miner system. Their implementation is based on representation of analyzed data by suitable strings of bits [19, 24]. The used way of implementation led to development of very fine tools to define the sets of relevant patterns. These tools bring the possibility to prepare input parameters tailor-made to answer the given analytical question. Note that an example is in Sect. 5.

- There are lot of items of background knowledge that are both good understandable for domain experts and enough formalized to be used in automatic generation of local analytical questions. Such automatically generated local analytical questions are also good understandable and moreover they are interesting from a point of view of domain experts. Many of these automatically generated local analytical questions can be answered by applications of GUHA procedures implemented in the LISp-Miner system. Some examples are given in Sect. 4 and Sect. 5.

- For each type of local analytical questions there is a well defined structure of a local analytical report answering particular local analytical question of the given type. The structure of the report is given by the chapters and sections and it is tailor-made to the solved analytical questions. The structure is filled in by the patterns found by the mining procedures and by the text explaining the role of presented patterns in answering the given analytical question.

- The preparation of the analytical report for the given local analytical question is usually not a straightforward process. Some iterations are necessary when applying particular GUHA procedures. Among other it is possible to filter out consequences of some pieces of the stored background knowledge [22]. But the current experience shows that these iterations and also the remaining steps can be done automatically using additional formalized knowledge.

- The semantics of the resulting local analytical reports can be formally described by the structure of the report, by patterns found by the mining procedures and by suitable formal description of the roles of particular patterns. This is a formal structure that can be used to index the analytical report instead of the key words textual documents are usually indexed. Such index can be used to storing and dissemination of local analytical reports through Semantic Web [14, 17].

- The patterns mined by the GUHA procedures can be understood as formulas of special logical calculi studied e.g. in [3, 6, 15]. There are important partial results on logical properties of these calculi. Namely results concerning deduction rules can be applied in various ways in the SEWEBAR project. Thus the research of logical calculi formulas of which correspond to patterns mined by the GUHA procedures is an integral part of the SEWEBAR project.

- Availability of local analytical reports through Semantic Web opens a possibility of (semi)automated creation of global analytical reports. Each global analytical report uses several local analytical reports found using Semantic Web. The global analytical questions and reports can be treated in a similar way like the local analytical questions and reports. It means that the global analytical questions are automatically generated and corresponding global analytical reports can be automatically produced, indexed, disseminated and stored for further use. The SEWEBAR project also covers research of these possibilities.

It is clear that the SEWEBAR is rather long time research project. However it is based on long time developed, functioning and relatively largely used LISp-Miner system. The SEWEBAR project will be realized gradually by enhancing the LISp-Miner system. The first necessary step is to enhance the architecture of the LISp-Miner system [28] and to implement tools for storing, maintaining and application of suitable items of formalized background knowledge. A new part of the LISp-Miner called *LM KS* (*LM KnowledgeSource*) was implemented to ensure new requirements related to the SEWEBAR project. This paper presents main features of the *LM KS* and outlines its possibilities.

We use two medical data sets mentioned in Sect. 2. The LM KnowledgeSource is introduced in Sect. 3. Formulation of local analytical questions using knowledge stored in the *LM KS* is sketched in Sect. 4. Application of the GUHA procedure SD4ft-Miner to solve a local analytical question formulated using *LM KS* is described in Sect. 5. Possibilities of using of knowledge stored in the *LM KS* in applications of the SD4ft-Miner are discussed in Sect. 6. Sect. 7 contains some concluding remarks and discussion on additional steps in realization of the SEWEBAR project.

2 Data Sets STULONG and ADAMEK

We deal with two data sets concerning cardiology – STULONG and ADAMEK. We use them as examples of databases of hospitals mentioned in Fig. 1. Data set STULONG concerns *Longitudinal Study of Atherosclerosis Risk Factors* (http://euromise.vse.cz/challenge2004/) [1]. Data set consists of four data matrices, we deal with data matrix *Entry* only. It concerns 1 417 patients – men that have been examined at the beginning of the study. Each row of data matrix describes one patient. Data matrix has 64 columns corresponding to particular attributes – characteristics of patients.

The second data set called ADAMEK concerns preventive cardiology. There is a definition of a set of items to be used to describe cardiologic patients. This set is called *Minimum data model of cardiology patient* [29]. Corresponding methodology is applied in two out-patients departments (Prague and Čáslav) since 2 000, see also http://www.euromise.org/health/preventive_cardio. html. The application results in a data matrix concerning 1122 patients (9/2006). Data matrix has 180 columns corresponding to particular attributes.

---

[1] The study (STULONG) was realized at the 2nd Department of Medicine, 1st Faculty of Medicine of Charles University and Charles University Hospital, U nemocnice 2, Prague 2 (head. Prof. M. Aschermann, MD, SDr, FESC), under the supervision of Prof. F. Boudík, MD, ScD, with collaboration of M. Tomečková, MD, PhD and Ass. Prof. J. Bultas, MD, PhD. The data were transferred to the electronic form by the European Centre of Medical Informatics, Statistics and Epidemiology of Charles University and Academy of Sciences (head. Prof. RNDr. J. Zvárová, DrSc).

3 LM KnowledgeSource

There are various types of background knowledge related to datasets STULONG and ADAMEK and maintained in the LM KnowledgeSource. Important portion of knowledge is given by a structure of a set of attributes. In our case, the attributes can be divided into 26 groups (e.g *Social characteristics*), see Tab. 1. Each pair of these groups has no common attributes and union of all groups is the set of all attributes. We call these groups *basic groups of attributes*. They are defined in *Minimum data model of cardiology patient* [29].

Table 1 Basic groups of attributes for STULONG and ADAMEK data sets

attribute	STULONG	ADAMEK
basic group of attributes *Personal information* - 3 attributes total		
Age	natural number	
Sex	M / F	
Department	*Prague*	*Prague / Čáslav*
basic group of attributes *Social characteristics* - 4 attributes total		
Marital status	*single / married / widowed / divorced*	
Education	*basic / apprentice / secondary / university*	*basic / secondary / higher*
Lives alone	not used	*yes / no*
Responsibility in a job	*manager / partly independent / pensioner*	not used
basic group of attributes *Physical examination* - 7 attributes total		
Weight (kg)	rational numbers range $\langle 41.6; 187 \rangle$	rational numbers range $\langle 52; 133 \rangle$
Systolic blood pressure (mm Hg)	natural numbers from the range $\langle 70; 230 \rangle$ categories defined as intervals $\langle 70; 80 \rangle, \ldots, \langle 220; 230 \rangle$	
...	...	...

Some of attributes in Tab. 1 are actually meta-attributes. It means that they differ in particular datasets what concerns their possible values. An example is meta-attribute *Education* that has two realizations - the attribute *Education* in STULONG with categories (i.e. possible values) *basic*, *apprentice*, *secondary university* and the attribute *Education* in ADAMEK with categories *basic*, *secondary*, *higher*. To be consistent we can formally deal also with meta-attribute *Sex* even if its both realizations have the same categories etc. Important knowledge concerning mutual influence of attributes is actually related to meta-attributes, see below.

Three examples of combination of various realizations of particular meta-attributes in both data sets follow, all examples are taken from Tab. 1.

- The meta-attribute is nominal or ordinal, it is realized in both data sets and it has the same categories (i.e. possible values) in both data sets, e.g. *Sex*.
- The meta-attribute is nominal or ordinal, it is realized in both data sets, but the sets of its categories in STULONG and ADAMEK are different, e.g. *Education*.

- The meta-attribute is realized in only one of data sets STULONG and ADAMEK, e.g. *Lives alone* and *Responsibility in a job*.

Definitions of meta-attributes and attributes together with possible categories are stored and maintained in the LM KnowledgeSource. Information on basic groups of (meta-)attributes as defined in Tab. 1 is very important, it reflects the *Minimum data model of cardiology patient* [29]. The basic groups of (meta-)attributes are perceived by physicians as reasonable sets of attributes. Thus the definition of basic groups of attributes is also stored in the LM KnowledgeSource.

There are also additional important groups of (meta-)attributes. An example is the group *Cardiovascular risk factors* that contains e.g. attributes *Hypertension, Mother hypertension, Obesity, Smoking* etc. coming from several basic groups. Thus there are tools also for maintaining definitions of additional groups of (meta-)attributes in the LM KnowledgeSource. The information on groups of attributes is used e.g. in formulation of local analytical questions see Sect. 4.

The above mentioned information has close relation to ontologies. There is some experience concerning ontologies in data mining with LISp-Miner [27]. However integration of ontology - like knowledge into LISp-Miner architecture will be very probably more effective than using general ontologies. We can moreover maintain also additional types of knowledge in the LM KnowledgeSource. An example is information on mutual influence of meta-attributes. This knowledge is managed in the way outlined in Fig. 2.

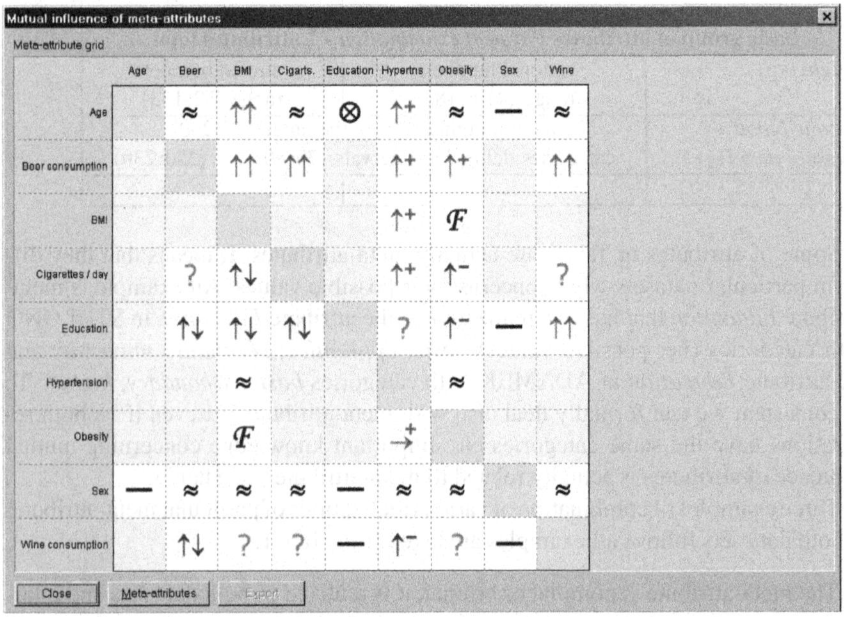

Fig. 2 Mutual influence of meta-attributes

There are four types of attributes: Boolean (e.g. Hypertension), nominal (e.g. *Sex, Department*), ordinal (e.g. Education), and cardinal (e.g. BMI i.e. Body Mass Index). There are several types of influences among attributes, most of them are relevant to specified types of attributes. Note that it is also possible to describe a conditional influence where some influence between two meta-attributes is observed only if some condition is met. The following types are used in Fig. 2:

- "–" – there is no influence
- $\otimes$ – there is some influence but we are not interested in
- $\approx$ – there is some influence
- $\uparrow\uparrow$ – if the row attribute increases then the column attribute increases too, both attributes are cardinal or ordinal
- $\uparrow\downarrow$ – if the row attribute increases then the column attribute decreases, both attributes are cardinal or ordinal
- $\uparrow^{+}$ – if the row attribute increases then the relative frequency of patients satisfying column attribute increases, row attribute is cardinal or ordinal and column attribute is Boolean
- $\uparrow^{-}$ – if the row attribute increases then the relative frequency of patients satisfying column attribute decreases, row attribute is cardinal or ordinal and column attribute is Boolean
- ? – there could be an influence, no detail is known
- $\mathscr{F}$ – means that there is a strong dependency like function, e.g. *obese* is equivalent to BMI ≥ 32
- $\rightarrow^{+}$ – truthfulness of the row attribute increases then relative frequency of true values of column attribute, both attributes are Boolean.

Note that there are also additional types and items of background knowledge stored and maintained in the LM KnowledgeSource. Two simple examples follow:

- basic value limits for particular attributes (e.g. 0 C and 100 C for temperature)
- typical interval lengths when categorizing values (e.g. 5 or 10 years for age).

4   Formulating Local Analytical Questions

There are various local analytical questions that can be generated on the basis of background knowledge [16]. The principle consists in application of a "local question pattern" to items of background knowledge. Examples of *items of background knowledge* are basic or additional groups of attributes or particular cells of Tab. 2 that are not marked by "–". Thus *Education* $\uparrow\downarrow$ *BMI* is an example of item of background knowledge. The signs like $\uparrow\uparrow$ or $\uparrow\downarrow$ are called *type of background knowledge*.

An example of a "local question pattern" (written in `typewriter`) is:

`Which items of background knowledge of the` type *type of background knowledge* `concerning` *attribute* `are observed in` *data set*?

An example of a local analytical question generated by this "local question pattern" is the question *Which items of background knowledge of the type* $\uparrow\downarrow$ *concerning attribute* Education *are observed in ADAMEK data?*

An additional "local question pattern" is the pattern :

What are the differences between particular values
of attribute *attribute* what concerns relation of Boolean
characteristics of groups of attributes *group 1 of attributes*
and *group 2 of attributes* in *data set* ?

It can be written in a symbolical way as

$$data\ set\text{:}\quad \bowtie^? (attribute)\text{:}\quad \mathcal{B}[group\ 1\ of\ attributes] \approx \mathcal{B}[group\ 2\ of\ attributes]$$

An example of such a local analytical question is the question

$$ADAMEK\text{:}\quad \bowtie^? (department)\text{:}\quad \mathcal{B}[Social\ characteristics] \approx \mathcal{B}[Alcohol]$$

that means *What are the differences between particular departments what concerns relation of Boolean characteristics of social characteristics and of alcohol drinking?* The attribute *department* has only two values - *Prague* and *Čáslav*, see Sect. 2. Thus we can write this analytical question in the form

$$ADAMEK\text{:}\quad Prague \bowtie^? Čáslav\ \text{:}\ \mathcal{B}[Social\ characteristics] \approx \mathcal{B}[Alcohol]\ .$$

This analytical question is solved in the next section.

There are lot of various "local question patterns" that can be applied to particular items of background knowledge stored in the LM KnowledgeSource. It is crucial that the local analytical questions generated this way can be solved by particular analytical procedures implemented in the LISp-Miner system. There are six analytical GUHA procedures in the LISp-Miner system that mines for various patterns verified on the basis of one or two contingency tables, [20, 16]. One of these procedures is the procedure SD4ft-Miner shortly described in the next section.

However the systematic description of all local question patterns is still a research task.

5   Applying SD4ft-Miner

We show how the local analytical question

$$ADAMEK\text{:}\quad Čáslav \bowtie^? Prague\ \text{:}\ \mathcal{B}[Social\ characteristics] \approx \mathcal{B}[Alcohol]$$

specified in the previous section can be solved by the SD4ft-Miner procedure. We use attributes - social characteristics *Marital status*, *Education* and *Lives alone*, (*Responsibility in a job* is not used in ADAMEK data) see Tab. 1. We also use three attributes describing alcohol consumption: *Beer*, *Wine*, and *Distillates*. The original values were transformed such that these attributes have the following possible values (i.e. categories):

Beer :   0, 1, ..., 7, 8–10, 11–15, > 15 [unit: 0.5 liter/week]
Wine :   0, 1, ..., 10, > 10 [unit: 0.2 liter/week]
Distillates :   0, 0.5, 1, 2, > 2 [unit: 0.05 liter/week].

The procedure *SD4ft-Miner* mines for SD4ft-patterns of the form

$$\alpha \bowtie \beta : \varphi \approx \psi / \gamma .$$

Here α, β, γ, φ, and ψ are Boolean attributes derived from the columns of analyzed data matrix $\mathcal{M}$ ($\mathcal{M}$ is ADAMEK in our case). The attributes α and β define two subsets of rows (i.e. subsets of patients in our case). The attribute γ defines a condition, it can be omitted in our case. The attributes φ and ψ are called *antecedent* and *succedent* respectively.

The SD4ft-pattern $\alpha \bowtie \beta : \varphi \approx \psi/\gamma$ means that the subsets of patients given by Boolean attributes α and β differ what concerns validity of association rule $\varphi \approx \psi$ when the condition given by Boolean attribute γ is satisfied. A measure of difference is defined by the symbol $\approx$ that is called *SD4ft-quantifier*. An example of an SD4ft-pattern is the pattern

$$\check{C}\acute{a}slav \bowtie Prague : \; married \Rightarrow^{D}_{0.228} Distillate(0) \wedge Wine(0) \, / \, male .$$

It means that the patients from Čáslav differ from the patients from Prague what concerns relation of Boolean attributes *married* and *Distillate*(0) $\wedge$ *Wine*(0) (i.e. does not drink neither distillates nor wine) when we consider only male patients.

The difference is given by the *SD4ft-quantifier* $\Rightarrow^{D}_{p}$ with parameter p, in our example it is $p = 0.228$. We introduce it using general notation α, β, γ, φ, and ψ for Boolean attributes derived from the columns of general analyzed data matrix $\mathcal{M}$. The *SD4ft-quantifier* concerns two four-fold contingency tables (i.e. 4ft-tables) $4ft(\varphi, \psi, \mathcal{M}/(\alpha \wedge \gamma))$ and $4ft(\varphi, \psi, \mathcal{M}/(\beta \wedge \gamma))$, see Fig. 3.

$\mathcal{M}/(\alpha \wedge \gamma)$	ψ	$\neg\psi$
φ	$a_{\alpha\wedge\gamma}$	$b_{\alpha\wedge\gamma}$
$\neg\varphi$	$c_{\alpha\wedge\gamma}$	$d_{\alpha\wedge\gamma}$

$4ft(\varphi, \psi, \mathcal{M}/(\alpha \wedge \gamma))$

$\mathcal{M}/(\beta \wedge \gamma)$	ψ	$\neg\psi$
φ	$a_{\beta\wedge\gamma}$	$b_{\beta\wedge\gamma}$
$\neg\varphi$	$c_{\beta\wedge\gamma}$	$d_{\beta\wedge\gamma}$

$4ft(\varphi, \psi, \mathcal{M}/(\beta \wedge \gamma))$

Fig. 3 4ft-tables $4ft(\varphi, \psi, \mathcal{M}/(\alpha \wedge \gamma))$ and $4ft(\varphi, \psi, \mathcal{M}/(\beta \wedge \gamma))$

The 4ft-table $4ft(\varphi, \psi, \mathcal{M}/(\alpha \wedge \gamma))$ of φ and ψ on $\mathcal{M}/(\alpha \wedge \gamma)$ is the contingency table of φ and ψ on $\mathcal{M}/(\alpha \wedge \gamma)$. The data matrix $\mathcal{M}/(\alpha \wedge \gamma)$ is a data sub-matrix of $\mathcal{M}$ that consists of exactly all rows of $\mathcal{M}$ satisfying $\alpha \wedge \gamma$. It means that $\mathcal{M}/(\alpha \wedge \gamma)$ corresponds to all objects (i.e. rows) from the set defined by α that satisfy the condition γ.

It is $4ft(\varphi, \psi, \mathcal{M}/(\alpha \wedge \gamma)) = \langle a_{\alpha\wedge\gamma}, b_{\alpha\wedge\gamma}, c_{\alpha\wedge\gamma}, d_{\alpha\wedge\gamma}\rangle$ where $a_{\alpha\wedge\gamma}$ is the number of rows of data matrix $\mathcal{M}/(\alpha \wedge \gamma)$ satisfying both φ and ψ, $b_{\alpha\wedge\gamma}$ is the number of rows of $\mathcal{M}/(\alpha \wedge \gamma)$ satisfying φ and not satisfying ψ, etc. The 4ft-table $4ft(\varphi, \psi, \mathcal{M}/(\beta \wedge \gamma))$ of φ and ψ on $\mathcal{M}/(\beta \wedge \gamma)$ is defined analogously.

The *SD4ft-quantifier* $\Rightarrow_p^D$ is related to the condition

$$\left| \frac{a_{\alpha \wedge \gamma}}{a_{\alpha \wedge \gamma} + b_{\alpha \wedge \gamma}} - \frac{a_{\beta \wedge \gamma}}{a_{\beta \wedge \gamma} + b_{\beta \wedge \gamma}} \right| \geq p .$$

This condition means that the absolute value of difference between the confidence of the association rule $\varphi \approx \psi$ on data matrix $\mathcal{M}/(\alpha \wedge \gamma))$ and the confidence of this association rule on data matrix $\mathcal{M}/(\beta \wedge \gamma))$ is at least p. The SD4ft-pattern $\alpha \bowtie \beta : \varphi \Rightarrow_p^D \psi / \gamma$ is *true on data matrix* $\mathcal{M}$ if the condition related to $\Rightarrow_p^D$ is satisfied on data matrix $\mathcal{M}$.

The SD-4ft pattern

$$\check{C}\acute{a}slav \bowtie Prague : married \Rightarrow_{0.228}^D Distillate(0) \wedge Wine(0) / male$$

is verified using the 4ft-tables $\mathcal{T}_{\check{C}\acute{a}slav}$ and $\mathcal{T}_{Prague}$ see Fig. 4 and Fig. 5. It is easy to verify that this pattern is true in ADAMEK data. Let us note that the sum of all frequencies from 4ft-tables $\mathcal{T}_B$ and $\mathcal{T}_C$ is smaller than 1122 because of omitting missing values.

The input of the procedure SD4ft-Miner consists of the SD4ft-quantifier and of several parameters that define the sets of relevant Boolean attributes α, β, γ, φ, and ψ. Their detailed description is out of the scope of this paper, see e.g. [19, 20].

ADAMEK / ($\check{C}\acute{a}slav \wedge$ male)	$Distillate(0) \wedge Wine(0)$	$\neg (Distillate(0) \wedge Wine(0))$
married	115	115
$\neg$ married	23	30

$\mathcal{T}_{\check{C}\acute{a}slav}$ = 4ft(*married, Distillate*(0) $\wedge$ *Wine*(0), ADAMEK /($\check{C}\acute{a}slav \wedge$ male))

ADAMEK / ($Prague \wedge$ male)	$Distillate(0) \wedge Wine(0)$	$\neg (Distillate(0) \wedge Wine(0))$
married	34	91
$\neg$ married	16	47

$\mathcal{T}_{Prague}$ = 4ft(*married, Distillate*(0) $\wedge$ *Wine*(0), ADAMEK /(*Prague* $\wedge$ male))

Fig. 4 4ft-tables $\mathcal{T}_{\check{C}\acute{a}slav}$ and $\mathcal{T}_{Prague}$

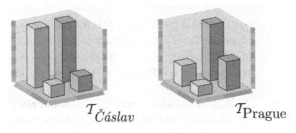

$T_{\check{C}\acute{a}slav}$ T_{Prague}

Fig. 5 4ft-tables $\mathcal{T}_{\check{C}\acute{a}slav}$ and $\mathcal{T}_{Prague}$ - graphical form

We show only possibilities suitable for our simple example. We do not use the condition γ, thus it is $a_{\alpha\wedge\gamma} = a_\alpha, b_{\alpha\wedge\gamma} = b_\alpha$, etc. There are tens of SD4ft-quantifiers, we use the above defined SD4ft-quantifier $\Rightarrow^D_{0.2}$ (i.e. $\Rightarrow^D_p$ for $p = 0.2$) together with two additional conditions:

$$\mid \frac{a_\alpha}{a_\alpha + b_\alpha} - \frac{a_\beta}{a_\beta + b_\beta} \mid \geq 0.2 \wedge a_\alpha \geq 30 \wedge a_\beta \geq 30 .$$

The set of relevant Boolean attributes α consists of one Boolean attribute *Department*(*Prague*), similarly for β and *Department*(Čáslav). The expressions *Department*(*Prague*) and *Department*(Čáslav) are examples of *basic Boolean attributes*. The basic Boolean attribute is an expression $A(\omega)$ where $\omega \subset \{a_1, \dots a_k\}$ and $\{a_1, \dots a_k\}$ is the set of all possible values of the attribute A.

The basic Boolean attribute $A(\omega)$ is true in the row o of data matrix in question if it is $a \in \omega$ where a is the value of A in the row o. The subset ω is called a *coefficient* of basic Boolean attribute $A(\alpha)$. The basic Boolean attribute *Department*(*Prague*) means that the patient was examined in Prague.

Let us remember that we are solving the local analytical question

ADAMEK: *Čáslav* $\bowtie^?$ *Prague* : $\mathcal{B}[Social\ characteristics] \approx \mathcal{B}[Alcohol]$.

The set $\mathcal{B}[Alcohol]$ of relevant succedents – Boolean attributes describing alcohol consumption is defined in Fig. 6. In Fig. 6 there is the expression

```
Min. length: 1    Max. length: 3
Literals boolean operation type: Conjunction
```

that means that each succedent is a conjunction of $1 - 3$ basic Boolean attributes chosen from the sets $\mathcal{B}[Beer]$, $\mathcal{B}[Wine]$, and $\mathcal{B}[Distillates]$ (maximally one attribute is used from each of these sets). In the row of attribute *Beer* in Fig. 6 there is the coefficient type defined as Interval with Length 1-6. It means that the coefficients ω of basic Boolean attributes *Beer*(ω) are intervals consisting of 1–6 categories of all 11 categories $\{ 0, 1, \dots, 7, 8\text{--}10, 11\text{--}15, > 15 \}$ of the attribute *Beer*, see above.

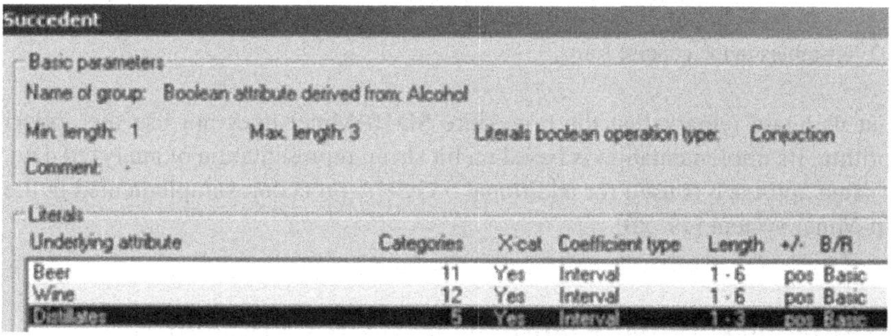

Fig. 6 Definition of the set $\mathcal{B}[Alcohol]$

Interval of length 3 consists of 3 *consecutive* categories. *Beer*(0,1,2) and *Beer*(1,2,3) are examples of basic Boolean attributes with coefficients – intervals of length 3. Similarly for additional lengths of intervals. The coefficient of the basic Boolean attribute *Beer*(1,4,9) is not the interval. We write *Beer*(0–2) instead of *Beer*(0,1,2), *Beer*(6–15) instead of *Beer*(6,7,8–10,11–15) etc. It is easy to verify that there are 51 basic Boolean attributes in the set $\mathcal{B}[Beer]$. Each of these basic Boolean attributes is of the form *Beer*(ω) where ω is an interval of the length 1–6. Similarly there are 57 basic Boolean attributes in the set $\mathcal{B}[Wine]$ (12 categories, coefficients specified as `Interval`, `Length 1-6` and 12 basic Boolean attributes in the set $\mathcal{B}[Distillates]$ (5 categories, coefficients specified as `Interval`, `Length 1-3`). It means that there are more than 38 000 basic Boolean attributes in the set $\mathcal{B}[Alcohol]$.

We define the set $\mathcal{B}[Social\ characteristics]$ in a similar way as conjunction of 1–3 basic Boolean attributes. However basic Boolean attributes with coefficients with one category are used together with two additional basic Boolean attributes *Education*(basic,secondary) and *Education*(secondary, higher). It means that there are more than $3*10^6$ relevant patterns $\check{C}aslav \bowtie Prague : \varphi \Rightarrow^D_{0.2} \psi$ such that $\varphi \in \mathcal{B}[Social\ characteristics]$ and $\psi \in \mathcal{B}[Alcohol]$. The procedure SD4ft-Miner generates and verifies them in 53 sec at PC with 1.6 GHz and 2GB RAM. There are 66 patterns satisfying given conditions, the strongest (i.e. with highest difference of confidences) is the pattern

$$\check{C}aslav \bowtie Prague : Marital\ status(married) \wedge Education(higher) \Rightarrow^D_{0.24} Beer(0)\ .$$

It says that the rule *Marital status*(*married*) $\wedge$ *Education*(*higher*) $\approx$ *Beer*(0) has confidence 0.24 higher for patients from *Čáslav* than for patients from *Prague*. The corresponding 4ft-tables (concise form) are in Fig. 7.

Čáslav	*Beer*(0)	$\neg$ *Beer*(0)
married $\wedge$ *higher*	39	28
$\neg$(*married* $\wedge$ *higher*)	272	315

Prague	*Beer*(0)	$\neg$ *Beer*(0)
married $\wedge$ *higher*	37	73
$\neg$(*married* $\wedge$ *higher*)	151	208

4ft(*married* $\wedge$ *higher*, *Beer*(0), *Čáslav*)   4ft(*married* $\wedge$ *higher*, *Beer*(0), *Prague*)

Fig. 7 4ft-tables in the concise form

Let us again remark that the procedure SD4ft-Miner does not use the apriori algorithm. Its implementation is based on bit string representation of analyzed data. The same approach is used for additional 5 GUHA procedures implemented in the LISp-Miner system [19, 20].

6   Using Background Knowledge in GUHA Procedures

GUHA procedures implemented in the LISp-Miner system are usually applied in an iterative way outlined in Fig. 8. It is supposed that both the minimal and the

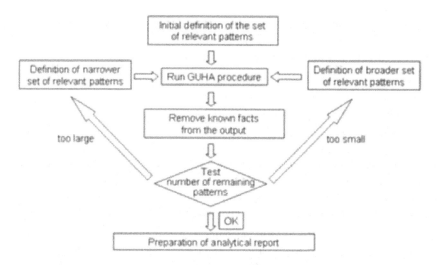

Fig. 8 Iterative application of the GUHA procedures

maximal number of patterns to be produced by the procedure are given. The main steps of the iterative process are:

- Initial definition of the set of relevant patterns
- Run of the GUHA procedure
- Removing known facts from the output of the GUHA procedure
- Test of the number of remaining patterns
- Definition of narrower set of relevant patterns
- Definition of broader set of relevant patterns

All these steps (of course except *Test of the number of remaining patterns*) require lot of knowledge. Both background knowledge related to the domain of application and knowledge related to the LISp-Miner is required. There is a research activity called *EverMiner* outlined in [20]. Its goal is to build a system of typical tasks, scenarios, repositories of antecedents and succedents, and additional tools to facilitate applications of particular GUHA procedures implemented in the LISp-Miner system. The final goal is to automatize the whole process outlined in Fig. 8. We only outline how the background knowledge can be used in this process.

All the GUHA procedures implemented in the LISp-Miner system use the definitions of the set of Boolean attributes to be automatically generated [19, 20]. A simple example of such definition is the definition of the set $\mathscr{B}[Alcohol]$ in Sect. 5. There are very fine tools to tune these definitions. For example, possibilities of definition of the set of coefficients (see e.g the term `Interval Length 1-6` for coefficients ω of basic Boolean attributes $Beer(\omega)$ in Fig. 6) must me combined with possibilities of definition of the set of particular categories (there are e.g. 11 categories { 0, 1, ..., 7, 8–10, 11–15, > 15 } of attribute *Beer*). To deal with all these tools effectively, it is necessary to use detailed information on attributes and their values.

Our goal is to produce the analytical reports automatically. It requires both automatic setting up of input parameters and automatic modification of input parameters of analytical procedures depending on results of mining, see steps

- Initial definition of the set of relevant patterns
- Definition of narrower set of relevant patterns
- Definition of broader set of relevant patterns

in Fig. 8. These steps can be divided into tens of particular relatively small tasks that can be solved automatically, however lot of relevant background knowledge is needed. Important portion of this knowledge is now stored in the LM Knowledge-Source as details concerning particular meta-attributes. In some cases it is related to particular attributes. The detailed description of this knowledge and of its application is out of the scope of this paper, we give only three examples.

- For attributes - instances of meta-attribute *Age* it is suitable to use intervals $\langle 0; 10 \rangle, \langle 10; 20 \rangle, \ldots$ or $\langle 0; 5 \rangle, \langle 5; 10 \rangle, \ldots$. The second possibility is used when the first one gives only few or no results.
- For attributes - instances of meta-attribute *Systolic blood pressure* it is suitable to use intervals $\langle 70; 80 \rangle, \langle 80; 90 \rangle, \ldots$ or $\langle 70; 75 \rangle, \langle 75; 80 \rangle, \ldots$. The second possibility is used when the first one gives only few or no results.
- A set of coefficients that will be automatically generated can determined using the rules (see also section 5):

  - *If attribute is ordinal then use* `Interval Length 1-W` *where*
    $W = \dfrac{\text{number of categories}}{3}$.
  - *If there are few or no results for* $W = \dfrac{\text{number of categories}}{3}$
    then use $W = \dfrac{\text{number of categories}}{2}$.

The step *Removing known facts from the output of the GUHA procedure* mentioned in Fig. 8 requires also background knowledge. However the more detailed explanation is out of the scope of this paper. A simple example is outlined in [22].

7 Conclusions and Further Work

We outlined the project SEWEBAR and we also shown that various forms of background knowledge stored in the LM KnowledgeSource can be used in formulation of local analytical questions. We demonstrated that one of reasonable local analytical questions formulated this way can be solved using GUHA procedure SD4ft-Miner that is a part of the LISp-Miner system. We outlined that background knowledge stored in the LM KnowledgeSource can be used to solve such questions automatically.

It follows also from additional experience [20, 24, 16, 27, 22] that the background knowledge stored in the LM KnowledgeSource can be used both to formulate and to solve additional local analytical questions. A way to produce resulting analytical report is outlined in [16], see also [26].

We can however also conclude that it is to early for a deeper evaluation of the presented approach. We suppose that its main effect will be in a possibility to generate a system of reasonable local analytical questions together with tools for answering each of formulated question. The correct evaluation requires evaluation of the whole system of questions.

Thus we will concentrate in our further work to develop software supporting formulation and answering local analytical questions as well as producing of corresponding analytical report. The experience from this process will be used in formulation and answering global analytical reports. The current experience makes possible to define particular research steps. We suppose to solve the following research tasks.

- Definition of a system of local analytical questions that can be formulated on the basis of background knowledge currently maintained in LM KnowledgeSource. We suppose to deal also with local analytical questions requiring application of several GUHA procedures.
- Development of a software system that will offer to a not-specialist in data mining particular local analytical questions in an intuitive way.
- Development of theory and software for filtering out of known facts from the output of particular GUHA procedures.
- Development of software modules that will assist to a user of LISp-Miner to solve defined local analytical questions. This research task can be solved in several phases:

    - The first phase will produce simple modules for initial definition of the set of relevant patterns for one selected GUHA procedure, see Fig. 8.
    - The last phase will finish by implementation of one complex module realizing the whole iterative process outlined in Fig. 8, including filtering out the known facts from the output of particular GUHA procedures. Note that the Fig. 8 must be modified to application of several GUHA procedures to solve complex local analytical questions.

- Definition of structure of analytical reports answering local analytical questions of particular types. Development of software preparing analytical reports answering particular local analytical questions. It is supposed that the results will be presented mainly in the form of Internet documents. However it will be possible to print each such report in a suitable form.
- Research on general properties of analytical reports presenting results of data mining. Large results covering general requirements concerning report writing and argumentation must be considered, see e.g. [2, 25]. The results will be used in preparation of analytical reports.
- Research of observational calculi. There are observational calculi formulas of which correspond to patterns mined by GUHA procedures 4ft-Miner and SD4ft-Miner implemented in the LISp-Miner system. Results concerning these calculi can be used e.g. in filtering known facts from output of particular procedures or

in dealing with analytical reports. Similar results concerning additional GUHA procedures are necessary.

- Research of possibilities of applications of other forms of background knowledge in all phases of dealing with local analytical questions. There are some promising preliminary results concerning ontologies [27].
- The results concerning ontologies are related to the *Ferda* data mining system that also deals with GUHA method. This system is younger than the LISp-Miner system and it has several unique features [13, 12]. It is now developed in parallel with the LISp-Miner system. It is supposed to use the Ferda data mining system in research related to the SEWEBAR project.
- Research on global analytical reports. The research can be divided into similar partial research tasks:

  - system of global analytical questions
  - software system that will offer particular global analytical questions in an intuitive way
  - software modules that will assist to solve defined global analytical questions
  - structure of analytical reports answering global analytical questions of particular types
  - software preparing analytical reports answering particular local analytical questions
  - application of results of research on general properties of analytical reports presenting results of data mining.

Acknowledgements. The work described here has been supported by the grant 201/08/0802 of the Czech Science Foundation.

References

1. Aggraval, R., et al.: Fast Discovery of Association Rules. In: Fayyad, U.M., et al. (eds.) Advances in Knowledge Discovery and Data Mining, pp. 307–328. AAAI Press / The MIT Press (1996)
2. Besnard, P., Hunter, A.: Elements of Argumentation. The MIT Press, Cambridge (2008)
3. Chrz, M.: Transparent deduction rules for the GUHA procedures. Diploma thesis. Faculty of Mathematics and Physics, Charles University in Prague (2007)
4. Hájek, P., Havránek, T.: Mechanising Hypothesis Formation - Mathematical Foundations for a General Theory. Springer, Heidelberg (1978)
5. Higgins, J.P.T., Green, S. (eds.): Cochrane Handbook for Systematic Reviews of Interventions 4.2.6. In: The Cochrane Library, vol. (4). John Wiley & Sons, Ltd, Chichester (2006) (updated September 2006)
6. Kodym, J.: Classes of SD4ft-patterns. Diploma thesis. Faculty of Mathematics and Physics, Charles University in Prague (in Czech) (2007)
7. Lín, V., Rauch, J., Svátek, V.: Content–based Retrieval of Analytical Reports. In: Schroeder, M., Wagner, G. (eds.) Proceedings of the International Workshop on Rule Markup Languages for Business Rules on the Semantic Web: In conjunction with the First International Semantic Web Conference ISWC 2002, pp. 219–224 (2002)

8. Lín, V., Rauch, J., Svátek, V.: Analytic Reports from KDD: Integration into Semantic Web. In: Malyankar, R. (ed.) Poster proceedings, ISWC 2002, p. 38. University of Cagliari (2002)

9. Lín, V., Rauch, J., Svátek, V.: Mining and Querying in Association Rule Discovery. In: Klemettinen, M., Meo, R. (eds.) Proceedings of the First International Workshop on Inductive Databases, pp. 97–98. University of Helsinki, Helsinki (2002)

10. Mareš, R., et al.: User interfaces of the medical systems - the demonstration of the application for the data collection within the frame of the minimal data model of the cardiological patient. Cor et Vasa. Journal of the Czech Society of Cardiology 44(suppl. 4), 76 (2002) (in Czech)

11. Matheus, J., et al.: Selecting and Reporting What is Interesting: The KEFIR Application to Healthcare Data. In: Fayyad U.M., et al. Advances in Knowledge Discovery and Data Mining, pp. 495–515. AAAI Press / The MIT Press (1996)

12. Ralbovský, M.: Evaluation of GUHA Mining with Background Knowledge. In: Joint proceedings PriCKL + Web Mining 2.0 workshops of ECML/PKDD 2007 PKDD, Warsaw, pp. 85–96 (2007)

13. Ralbovský, M., Kuchař, T.: Using Disjunctions in Association Mining. In: Perner, P. (ed.) ICDM 2007. LNCS (LNAI), vol. 4597, pp. 339–351. Springer, Heidelberg (2007)

14. Rauch, J.: Logical Calculi for Knowledge Discovery in Databases. In: Komorowski, J., Żytkow, J.M. (eds.) PKDD 1997. LNCS, vol. 1263, pp. 47–57. Springer, Heidelberg (1997)

15. Rauch, J.: Logic of Association Rules. Applied Intelligence 22, 9–28 (2005)

16. Rauch, J.: Project SEWEBAR – Considerations on Semantic Web and Data Mining. In: Prasad, B. (ed.) Proceedings of 3rd Indian International Conference on Artificial Intelligence – IICAI 2007 [CD-ROM], pp. 1763–1782. Florida A&M University, Tallahassee (2007)

17. Rauch, J.: Observational Calculi – Tool for Semantic Web (Poster abstract). In: Sure, Y. (ed.) Proceedings of the Poster Track of the 5th European Semantic Web Conference, ESWC 2008 (2008), `http://sunsite.informatik.rwth-aachen.de/Publications/CEUR-WS/Vol-367/` (cited February 26,2009)

18. Rauch, J., Šimůnek, M.: Mining for 4ft Association Rules. In: Morishita, S., Arikawa, S. (eds.) DS 2000. LNCS, vol. 1967, pp. 268–272. Springer, Heidelberg (2000)

19. Rauch, J., Šimůnek, M.: An Alternative Approach to Mining Association Rules. In: Lin, T.Y., et al. (eds.) Data Mining: Foundations, Methods, and Applications, pp. 219–238. Springer, Heidelberg (2005)

20. Rauch, J., Šimůnek, M.: GUHA Method and Granular Computing. In: Hu, X., et al. (eds.) Proceedings of IEEE conference Granular Computing, pp. 630–635 (2005)

21. Rauch, J., Šimůnek, M.: Semantic Web Presentation of Analytical Reports from Data Mining – Preliminary Considerations. In: Lin, T.Y., et al. (eds.) Web Intelligence 2007 Proceedings, pp. 3–7 (2007)

22. Rauch, J., Šimůnek, M.: LAREDAM - Considerations on System of Local Analytical Reports from Data Mining. In: An, A., Matwin, S., Raś, Z.W., Ślęzak, D. (eds.) Foundations of Intelligent Systems. LNCS (LNAI), vol. 4994, pp. 143–149. Springer, Heidelberg (2008)

23. Rauch, J., Tomečková, M.: System of Analytical Questions and Reports on Mining in Health Data – a Case Study. In: Roth, J., et al. (eds.) Proceedings of IADIS European Conference Data Mining 2007, pp. 176–181. IADIS Press (2007)

24. Rauch, J., Šimůnek, M., Lín, V.: Mining for Patterns Based on Contingency Tables by KL-Miner – First Experience. In: Lin, T.Y., et al. (eds.) Foundations and Novel Approaches in Data Mining, pp. 155–167. Springer, Heidelberg (2005)

25. Shearring, H.A., Christian, B.C.: Reports and how to write them, 147 p. Georg Allen and Unwin Ltd, London (1967)
26. Strossa, P., Černý, Z., Rauch, J.: Reporting Data Mining Results in a Natural Language. In: Lin, T.Y., et al. (eds.) Foundations of Data Mining and Knowledge Discovery, pp. 347–362. Springer, Heidelberg (2005)
27. Svátek, V., Rauch, J., Ralbovský, M.: Ontology-Enhanced Association Mining. In: Ackermann, M., Berendt, B., Grobelnik, M., Hotho, A., Mladenič, D., Semeraro, G., Spiliopoulou, M., Stumme, G., Svátek, V., van Someren, M. (eds.) EWMF 2005 and KDO 2005. LNCS, vol. 4289, pp. 163–179. Springer, Heidelberg (2006)
28. Šimůnek, M.: Academic KDD Project LISp-Miner. In: Abraham, A., et al. (eds.) Advances in Soft Computing – Intelligent Systems Design and Applications, pp. 263–272. Springer, Heidelberg (2003)
29. Tomečková, M.: Minimal data model of the cardiological patient - the selection of data. Cor et Vasa. Journal of the Czech Society of Cardiology 44(suppl. 4), 123 (2002) (in Czech)

Part II
Web Mining 2.0

Item Weighting Techniques for Collaborative Filtering

Linas Baltrunas and Francesco Ricci

Abstract. Collaborative Filtering (CF) recommender systems generate rating predictions for a target user by exploiting the ratings of similar users. Therefore, the computation of user-to-user similarity is an important element in CF; it is used in the neighborhood formation and rating prediction steps. In this paper we investigate the role of item weighting techniques. An item weight provides a measure of the importance of an item for predicting the rating of another item and it is computed as a correlation coefficient between the two items' rating vectors. In this paper we analyze a wide range of item weighting schemas. Moreover, we introduce an item filtering approach, based on item weighting, that works by discarding in the user-to-user similarity computation the items with the smallest weights. We assume that the items with smallest weights are the least useful for generating the prediction. We have evaluated the proposed methods using two datasets (MovieLens and Yahoo!) and identified the conditions for their best application in CF.

1 Introduction

Recommender systems are powerful tools helping on-line users to tame information overload and providing personalized recommendations on various types of products and services [20]. Collaborative Filtering (CF) is a successful recommendation technique which automates the so-called "word-of-mouth" social strategy. When generating a recommendation for a target user, CF analyzes the experiences/evaluations of similar users that are modeled as vectors of item ratings, and it recommends the items that the similar users liked the most [24]. The user-to-user similarity is usually determined by a pair-wise comparison of *all* the available ratings explicitly expressed by the users about items in the product catalog. The assumption is that two highly

Linas Baltrunas and Francesco Ricci
Free University of Bozen-Bolzano
Domeninkanerplatz 3, Bozen, Italy
e-mail: {lbaltrunas,fricci}@unibz.it

B. Berendt et al. (Eds.): Knowl. Disc. Enhan. with Sem. and Soc. Info., SCI 220, pp. 109–126.
springerlink.com

correlated users will have similar tastes also on items they have not rated yet. It is therefore evident that user-to-user similarity is a core computation step in CF [2].

Note that in this paper we are focusing on user-to-user CF, as opposed to item-to-item CF. The latter computes the rating prediction for a target user and item pair as a weighted average of the ratings the user gave to *similar items*, and therefore user-to-user similarity is not exploited.

In CF, user-to-user similarity computation is used in the neighborhood formation and in the final rating prediction. As we noted above, two users are considered similar if they have expressed similar opinions (ratings) on a set of co-rated items. In standard CF each of the co-rated items contributes equally to the similarity. But it seems natural to conjecture that some items could be more important than others when computing the user-to-user similarity. For example, if the movie "Dead on Arrival" gets the lowest possible rating from all the users, then it should not be very useful in understanding whether two users have similar preferences. These type of items, i.e., with very similar ratings in the users' population, should contribute less than others or could be even discarded from the similarity computation. Moreover, additional information about the *content* of the items could potentially improve the performance of CF. For example, consider two users who have very similar tastes for action movies but very different tastes for dramas and documentary movies. In general, these users would have a small correlation when comparing all their co-rated items. However, when considering only the action movies the correlation would be higher. When predicting a rating for an action movie, co-rated action movies could be given a higher importance (weight).

In this paper we propose to address the previously mentioned issues by using item weighting, i.e., by extending the user-to-user similarity definition so that it is possible to *weight* the influence of each item. This is not an easy task. There are many sources of information potentially useful for computation of item weights. Moreover, it is not clear how this information can be transferred into item weights. For instance, it is questionable whether it would be better to consider rating data statistics, or to exploit information coming from external sources, such as item descriptors or contextual information. We also need to determine how strongly the additional information should influence the similarity. Finally, we observe that after having found good information sources (for determining the importance of items), there is no well-accepted way (formula) to extend the standard user-to-user correlation measure in order to take into account these weights.

In this paper we address the issues mentioned above, and in particular:

- We analyze a wide range of weighting schemas — some new and some already used in previous research. These are based on various types of information; one group of methods explores the data dependencies between item ratings, whereas another tries to exploit the information contained in the descriptions of the items.
- We empirically investigate three approaches to incorporate the computed weights into the user-to-user correlation measure and we provide an analysis of their behavior. We also conduct some experiments aimed at understanding to what extent the additional information contained in the weights should modify the prediction formula, compared to the standard similarity metric.

- Additionally, we analyze the behavior of an item filtering method based on the computed weights. This study evaluates wether it is possible to improve the prediction accuracy by using just a small subset of items, i.e., those that are very relevant to the target item.

In this paper we show that item weighting can improve the accuracy of the baseline CF (for both datasets), for all the neighborhood sizes, and for all the error measures we used. This improvement is obtained by a particular weighting method based on Singular Value Decomposition (SVD-PCC) of the rating matrix. The other weighting schemas did not uniformly improve the baseline, however, in some situations they can even produce larger improvements over the baseline than SVD-PCC. We also show that the largest reduction of the mean absolute error was achieved by using item filtering together with a particular weighting method but at the cost of sacrificing coverage. We also identified a weight incorporation method that uniformly performs better than the others.

The rest of the paper is structured as follows. Section 2 gives an overview of the state of the art in item weighting and item filtering. In Section 3 we introduce our definition of a user-to-user similarity that incorporates item weighting. There, we explain and motivate different weighting schemas (Subsection 3.1), and different approaches to integrate the weights into the similarity measure (Subsection 3.2). In Section 4 we introduce our new approach to item filtering. Section 5 provides the experimental evaluation and the discussion of proposed methods. Section 6 summarizes the work presented in this paper and draws some conclusions.

2 Related Work

CF can be seen as an instance-based learning approach where users are instances described by their feature vectors, and the product (item) ratings play the role of feature values. Hence, the rationale for the application of feature weighting/selection techniques in CF is that some items may be more important than others in the similarity computation [8].

Feature weighting is a well studied problem in Machine Learning, and in Instance-Based Learning in particular [17]. A number of studies reported accuracy improvements when features are weighted in the instance-to-instance distance computation ([26, 3] offer an excellent survey). In the context of CF feature weighting can be called item weighting, since features of a user are actually item ratings. The huge amount of items and the data sparsity makes most of these classical methods inefficient.

Item weighting and item selection have not been widely studied in CF, and only few researchers tried to apply these methods. In [8], the authors adapted the idea of inverse document frequency, a popular Information Retrieval measure [21], to item weighting in CF. The key idea of their approach, which is called Inverse User Frequency, is that universally liked and known items do not give much information about the true user preferences. Therefore, the weights for such commonly rated items should be decreased. This approach showed better performances than the *unweighted* version. The same idea of decreasing the weights of commonly liked items was implemented by using variance weighting [12]. Here, items with bigger variance are supposed to distinguish better the taste of the users and, therefore, they receive larger weights in the user-to-user similarity formula. Yet their results were not encouraging and the authors report just a small increase of the Mean Absolute Error (MAE) with respect to the non-weighted version.

In [28] Information Theoretical approaches for item weighting were introduced. The authors showed that Mutual Information and Entropy based methods perform better than Inverse User Frequency and standard CF. Moreover, Mutual Information obtained a 4.5% accuracy improvement and performed better (even when trained on a small number of instances) than the standard CF. They also reported that Inverse User Frequency, differently from [8], decreased the accuracy. [14] presents another automatic weighting schema for CF systems. This method tries to find weights for different items that bring each user even closer to the similar users and farther from dissimilar users. That method uses ideas from model-based approaches and can decrease MAE. As it is clear from this short review, the quoted papers offer contradicting results. For this reason we decided to test ourselves the performance of some of these methods on two different data sets.

In addition to feature weighting, feature selection (filtering) algorithms have also been very popular in Machine Learning [26, 15, 16], and they are now receiving new interest because of their exploitation in Information Retrieval ranking methods based on learning [19, 10]. In CF, there have been just a couple of attempts to use item selection. An interesting approach to item selection was proposed in [1]. To generate a recommendation the authors proposed to use only the items (ratings) that were experienced in the same context of the target item prediction. The authors made item selection only in the subset of the ratings where cross-validation showed improvement in the accuracy. But, due to sparsity and small size of their data, only a small improvement over the baseline CF method was obtained.

In our previous work [4] we showed that the straightforward application of item selection methods based on the *filter* approach [26] fails because of the high sparsity of the data. Besides, *wrapper* methods seem not to be applicable because of the large number of features and therefore the large complexity of evaluating the exponential number of possible subsets of the features. Therefore, in [5] we developed a new filter-based approach that tries to overcome these problems. The item selection is made according to a pre-computed set of weights and considers only the items that are present in the profiles of both users whose similarity is sought. That approach improved all the error measures we used [5]. In this paper we evaluate a simpler approach that does not depend on the pair of users whose similarity is sought. While computing user-to-user similarity It simply tries to discard the items with lowest weights.

3 Weighted Similarity

In order to use item weighting in user-to-user similarity we first compute the item weights using a weighting schema, and then we incorporate these weights into the standard user-to-user similarity measure. To compute the item weights we use a wide range of weighting schemas that are described in Subsection 3.1. Each weighting schema computes a real valued $n \times n$ weight matrix W (n is the number of items). An entry w_{ij} is the weight (importance) of item i for predicting the ratings of item j (for all the target users). In order to be able to compare the behavior of different weighting schemas, in this paper we normalized the weights to range in $[0, 1]$.

The role of the target item j needs some further explanation. In fact this gives to an item weighting schema the flexibility to assign to each predictive item a weight that depends on the target item whose rating predictions are sought. Hence, the weights used for predicting the ratings for item j (for all the users) can be different from those used in the predictions for another item j'. In this way, we encode in a weight how much two items are correlated, and what role an item should play when a prediction is sought for the second one. Without such a flexibility we could only express more general knowledge about the importance of an item for all the rating predictions. In fact, an example of this approach is provided by the variance weighting method that is explained in Subsection 3.1. Finally, given a weighting schema, there are several ways to integrate the weights into an (unweighted) similarity measure. These are methods described in Subsection 3.2.

3.1 Weighting Schemas

Item weighting tries to estimate how much a particular item is important for computing the rating predictions for a target item. In CF, item weights can be learned while exploring training data consisting of user ratings or using additional information associated with the items, i.e., their descriptions. Based on this observation we can partition weighting methods in two groups. The methods in the first group determine the item weights using statistical approaches and are aimed at estimating statistical properties of the ratings and the dependencies between items. For instance, the variance of the item ratings (provided by a users' population) belongs to the first group of methods. The second group of methods uses item descriptions to estimate item dependencies and finally infer the weights. For instance, the similarity between item descriptions can be used to infer item-to-item dependencies and these can be converted into weights.

In the rest of this paper we will consider the following item weighting approaches: Random, Variance, Mutual Information, Genre, Pearson Correlation Coefficient (PCC), and PCC computed on low dimensional item representations. Let us now precisely define these methods.

Random. The first method is used as a reference for comparing different approaches. It assigns to each predictive (i) and target (j) item combination a random weight sampled from the uniform distribution in $[0, 1]$: $w_{ij} = random([0, 1])$

Variance. The variance method is based on an observation originally made in [12]. It gives higher weights to items with higher variance among the ratings provided by the users to that item:

$$w_{ij} = \frac{\sum_u (v_{ui} - \bar{v}_i)^2}{\#users\ who\ rated\ i}$$

Here the sum runs over all the users who rated the item i, v_{ui} is the rating given by user u to item i, and $\bar{v}_i$ is the mean of the ratings provided by all the users to the item i. The variance feature weighting method uses only information on the item whose ratings have to be predicted. All the methods that we are presenting next, explore relations between predictive items and the target item.

IPCC. The first method in this group uses Pearson Correlation Coefficients (PCC) as a measure of the dependency between two vectors representing the ratings of all users for two items:

$$w_{ij} = \frac{\sum_u (v_{ui} - \bar{v}_i)(v_{uj} - \bar{v}_j)}{\sqrt{\sum_u (v_{ui} - \bar{v}_i)^2 \sum_u (v_{uj} - \bar{v}_j)^2}}$$

Here the index u runs over all the users that have rated both items i and j, and $\bar{v}_i$ is the mean of the ratings on item i. We observe that IPCC (Item Pearson Correlation Coefficients) builds a symmetric weight matrix W, i.e., the importance of the item i in the prediction of the ratings for the item j is equal to the importance of the item j when it is used to predict the ratings for the item i.

Mutual Information. Mutual Information (MI) measures the information that a random variable provides to the knowledge of another. In CF, following [28], we compute the mutual dependency between the target item variable and the predictive item variable (the values are the ratings). The Mutual Information of two random variables X and Y is defined as:

$$I(X;Y) = \sum_x \sum_y p(x,y) \log \frac{p(x,y)}{p(x)p(y)}$$

The above equation can be transformed into $I(X;Y) = H(X) + H(Y) - H(X,Y)$, where $H(X)$ denotes the entropy of X, and $H(X,Y)$ is the joint entropy of X and Y. Using the above formula Mutual Information computes the weights as follow:

$$w_{ij} = - \sum_{r=1}^{5} p(v_i = r) \log p(v_i = r) - \sum_{r=1}^{5} p(v_j = r) \log p(v_j = r)$$

$$+ \sum_{r'=1}^{5} \sum_{r''=1}^{5} p(v_i = r', v_i = r'') \log p(v_i = r', v_i = r'')$$

Here the probabilities are estimated with frequency counts in the training dataset, hence for instance $p(v_i = r) = \frac{\#users\ who\ rated\ i\ as\ r}{\#users\ who\ rated\ i}$ estimates the probability that the item i is rated with value r. We observe that r is running through all rating values, and in our data sets this is equal to 5.

SVD-PCC. The methods described previously, i.e., IPCC and Mutual Information, compute a form of statistical dependency between two items. Note that such statistics between two items are computed using the pairs of ratings that all the users expressed for *both* items. When the data is extremely sparse there can be only few users who have rated both items, and the computation of the statistical dependency based on this incomplete information could be inaccurate. Moreover, the fact that two items were not co-rated by the users should not imply that they are not similar [7]. In order to avoid such problems we decided to reduce the dimensionality of the ratings matrix and compute item-to-item correlation in the transformed space rather than on the original raw data. This approach is based on Latent Semantic Analysis (LSA) [9], originally used for analyzing the relationships between a set of documents and the contained terms. LSA is now widely applied to CF to generate accurate rating predictions, and the user-to-user similarity is computed in a space with lower dimension than the number of items and users [22]. LSA uses Singular Value Decomposition (SVD), a well-known matrix factorization technique (see [11] for a reference).

SVD factors a rating matrix $R = (v_{ij})$, of size $m \times n$ (m users and n items) into three matrixes as follows: $R_{m \times n} = U_{m \times r} \cdot S_{r \times r} \cdot V_{r \times n}^T$ where, U and V are two orthogonal matrices, r is the rank of the matrix R and S is a diagonal matrix having all the singular values of matrix R as its diagonal entries. Note, that all the singular values are positive and ordered in descending order. Selecting the $k < r$ largest singular values, and their corresponding singular vectors U and V, one gets the rank k approximation $R_{m \times n} = U_{m \times k} \cdot S_{k \times k} \cdot V_{k \times n}^T$ of the rating matrix R.

Each column of the matrix $I = V_{k \times n}^T$ gives a dense representation of the corresponding item in a k dimensional space. The choice of k is an open issue in the factor analytic literature [9] and it is typically application dependent. We set $k = 40$ in our experiments (see Section 5 for more details). To compute the similarity of item i and item j we compute the PCC between two columns of I:

$$ w_{ij} = \frac{\sum_k (I_{ki} - \bar{I}_i)(I_{kj} - \bar{I}_j)}{\sqrt{\sum_k (I_{ki} - \bar{I}_i)^2 \sum_k (I_{kj} - \bar{I}_j)^2}} $$

where the sum runs over the row index of the matrix I and $\bar{I}_j$ denotes the average value of the j-th column.

Genre Weighting. All the previous methods exploit statistics of the users' rating data to determine item weights. The last method that we present here computes the weights using descriptions of the items. In the movie recommendation data sets, which we are using for our experiments, the movies are tagged with movie genres. Hence, we make the assumption that the larger the number of common genres, the higher is the dependency. Hence, the weight of the predictive item i for predicting the ratings of the target item j is given by:

$$ w_{ij} = \frac{\#\ comon\ genres\ of\ items\ i\ and\ j}{\#genres} $$

Genre weighting is related to the methods presented in [6], where item description information is used to selectively choose the items to be used in the user-to-user similarity.

3.2 Weight Incorporation

The two most popular measures of user-to-user similarity are: the cosine of the angle, formed by the rating vector of the two users; and Pearson Correlation Coefficient (PCC). PCC is preferred when data contains only positive ratings, and has been shown to produce better results in such cases [8]. The PCC between the users x and y is defined as:

$$PCC(x,y) = \frac{\sum_i (v_{xi} - \bar{v}_x)(v_{yi} - \bar{v}_y)}{\sqrt{\sum_i (v_{xi} - \bar{v}_x)^2 \sum_i (v_{yi} - \bar{v}_y)^2}}$$

where v_{xi} denote the rating given by the user x to the item i, and $\bar{v}_x$ is the mean of the ratings of the user x. The sum runs over all the items i that both users have rated.

As we mentioned in the Introduction, the motivation for using weights in the user-to-user similarity is to improve the prediction accuracy. This approach has been used in instance-based learning to produce weighted the Euclidean metric [3]. We must observe that the effect of the weights on the Euclidean metric has a clear (geometric) interpretation: the features with larger weights produce a greater impact on the similarity. For PCC the situation is not as clear as for the Euclidean metric. Because of the more complicated nature of PCC, there exists no single well accepted approach to extend it with weights. In our study we analyzed three different versions of *weighted* PCC. The first version which we call normalized Weighted PCC ($WPCC_{norm}$) makes a natural extension of the unweighted PCC:

$$WPCC_{norm}(x,y,j) = \frac{\sum_i w_{ij}(v_{xi} - \bar{v}_x)(v_{yi} - \bar{v}_y)}{\sqrt{\sum_i w_{ij}(v_{xi} - \bar{v}_x)^2 \sum_i w_{ij}(v_{yi} - \bar{v}_y))^2}}$$

where j denotes the target item of the rating prediction, and w_{ij} is the weight of the (predictive) item i when making a prediction for j. Here $\bar{v}_x$ is the average rating of the user x.

The second approach we consider is used by SAS^{TM}[23] and it extends the previous basic approach by replacing the standard average of the user ratings by a weight-normalized version:

$$WPCC_{wavg}(x,y,j) = \frac{\sum_i w_{ij}(v_{xi} - \hat{v}_x)(v_{yi} - \hat{v}_y)}{\sqrt{\sum_i w_{ij}(v_{xi} - \hat{v}_x)^2 \sum_i w_{ij}(v_{yi} - \hat{v}_y)^2}}$$

where $\hat{v}_x = \frac{\sum_i w_{ij}x_i}{\sum w_{ij}}$ and we note that this new normalized average depends on the target item j. The idea here is that the *weighted* average of the user ratings should better estimate the user average. Note that we compute the average rating using all the ratings of the user, rather then just overlapping items.

The third method that we consider in this paper was used in [14], and it differs from the previous ones as it does not normalize the similarity metric:

$$WPCC_{no-norm}(x,y,j) = \frac{\sum_i w_{ij}(v_{xi} - \bar{v}_x)(v_{yi} - \bar{v}_y)}{\sqrt{\sum_i (v_{xi} - \bar{v}_x)^2 \sum_i (v_{yi} - \bar{v}_y)^2}}$$

In this approach, the users who co-rated highly weighted items would have higher correlation. Because of the lack of the normalization this correlation measure depends on the absolute values of weights, rather than on the relative differences between the weights. Therefore, the similarity that is computed on the items with small weights will be small.

Further discussion on the different weight incorporation methods can be found in the experimental part of this paper.

4   Item Selection

In item selection instead of using the precise values of the weights for tuning the similarity function, a simpler decision is taken to either use or not to use an item, based on the value of its weight. In fact, item selection could be seen as a particular case of item weighting, where just binary weights are used.

In fact, finding the optimal subset of items to be used in the user-to-user similarity would require an extensive search through the space of all the subsets of the items [15]. Applying this to a recommender system scenario would require to conduct a search procedure for every target item (the item playing the role of the class to be predicted) and for a large number of subsets of the predictive items. This is clearly extremely expensive, and therefore, we propose to use a more parsimonious approach (filter method) [15]. It uses the information provided by an item weighting method to select, for each target item an appropriate set of predictive items. Hence, first we compute the item weights using one of the weighting schemas described above, and then we select only the relevant items, i.e., the items with the largest weights, for any given target item.

Users tend to rate a small part of the items in a data set. Selecting a small number of items for the similarity computation further reduces the amount of information used to compute it. Therefore, in [5] we proposed to select the items that are present in the profiles of both users, whose similarity must be computed. We showed that this approach can increase the prediction accuracy (for many different error measures), and can use a smaller number of nearest neighbors.

In this paper we propose and evaluate a different approach. Instead of performing a dynamic item selection which depends on the target user we simply filter out (and never consider in the similarity computation) the items with the smallest weights.

In other words, if a weight is less than a given threshold, we set it to zero. Note, that instead of using binary weights as we did in [5], we still use in the user-to-user similarity computation the weights in $[0, 1]$. Moreover, the set of items used in a prediction depends only on the target item and it is the same for all the users.

5 Experimental Evaluation

In this section we evaluate the item weighting and item selection methods presented so far. We observe that to generate the prediction, one of the three variations of $WPCC$ was used in computing the nearest neighbors of the target users and also in the prediction step:

$$v_{xj}^* = \bar{v}_x + \frac{\sum_{y \in N(k,x,j)} WPCC(x,y,j) \times (v_{yj} - \bar{v}_y)}{\sum_{y \in N(k,x,j)} |WPCC(x,y,j)|}$$

Here the sum runs over the k-nearest neighbors $N(k,x,j)$ of the user x. Note that the neighbors depend on the target item j because the user-to-user similarity depends on the target item. In our implementation of CF, as previously done in other studies, when making a rating prediction for a user x we take into account only the neighbors with at least a minimum number of co-rated items with the target user (six in our case) [6].

In the experiments, we used two data sets with ratings in $\{1,2,3,4,5\}$.[1] The MovieLens [18] data set contains 100K ratings, for 1682 movies by 943 users, who have rated 20 and more items. The data sparsity of this dataset is 96%. The Yahoo! [27] Webscope movies data set contains 221K ratings, for 11915 movies by 7642 users. The data sparsity of this dataset is much higher, 99.7%. To evaluate the described methods we used holdout validation, where both datasets where divided into train (80%) and test (20%) subsets. We used the train data to learn the weights and also to generate the prediction for the test ratings. Because of the high computational complexity, when computing weights, we were not able to perform a multi-fold cross-validation.

To measure the accuracy we used Mean Absolute Error (MAE), coverage and F_1 (in the rest of the paper denoted as F) measures [13]. To compute F, we considered an item worth recommending only if the user rated it as 4 or 5. Coverage is the fraction of the ratings in the test set for which predictions can be made. Note that when computing MAE we followed the standard practice and did not take into account the ratings that an algorithm was not able to predict.

To compute Singular Value Decomposition we used the SVDLIB [25] library and selected $k = 40$ biggest singular values. As mentioned earlier, the selection of k is a problematic issue. As stated above, in our experimental setting the parameter selection using cross-validation was too expensive. Selecting a too small k can lead to a big loss of information, however, selecting too big k could result in not removing

---

[1] We want to thank both Grouplens and Yahoo! for making their data sets available.

the noise present in the data. Computing similarity in the higher dimensional space might also not be effective because of the "curse of dimensionality". Our selection was based on some trials and the analysis of the outcome of previous reports [9, 7].

5.1 Weight Incorporation

We started our experiments by evaluating the performance of the three weight incorporation methods described in Subsection 3.2. In a first experiment we used all the weighting schemas and computed F and MAE for the three weight incorporation methods. For lack of space we focus here only on SVD-PCC weighting. The comparison of three weight incorporation methods for the two considered datasets, using the *SVD-PCC* weighting schema, is depicted in Figure 1.

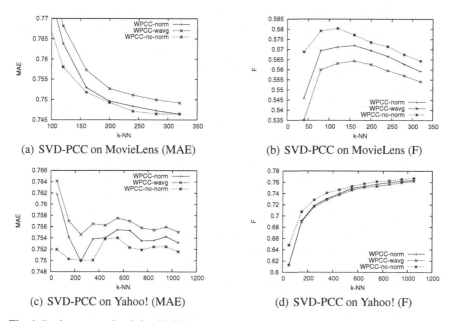

(a) SVD-PCC on MovieLens (MAE)

(b) SVD-PCC on MovieLens (F)

(c) SVD-PCC on Yahoo! (MAE)

(d) SVD-PCC on Yahoo! (F)

Fig. 1 Performance of weighted PCC measures

We observe here that most of the time $WPCC_{no-norm}$ performs better than the other methods independently of the data set, and independently of the error measure we used. Because of lack of space we do not show the results of this experiment with the other weighting schemas but they confirm this result, i.e., $WPCC_{no-norm}$ always performs better that the other twos.

This is quite surprising, as $WPCC_{norm}$ or $WPCC_{WAVG}$ are used in the majority of the literature and software packages [23, 28]. In fact, because of the complicated behavior of PCC such results are hard to interpret. Our explanation is that $WPCC_{no-norm}$ gives a higher absolute correlation to a user who has expressed ratings on more correlated items to a target item. For example, consider the following three users with ratings:

w_{ti}	0.8	0.7	0.6	0.5
	i_1	i_2	i_3	i_4
u_1	?	?	1	5
u_2	2	4	2	4
u_3	1	5	?	?

Here the first line of the table gives the item weights which are used in the similarity computation. Computing the user-to-user similarity on this toy example gives an insight into the nature of the different correlations. We have the following: $WPCC_{norm}(u_1,u_2) = 1, WPCC_{norm}(u_3,u_2) = 1, WPCC_{no-norm}(u_1,u_2) = 0.55, WPCC_{no-norm}(u_3,u_2) = 0.75$. One can see that $WPCC_{norm}$ takes into account only the relative size of the weights that are used in the similarity computation and not their absolute value. In fact, the similarity of u_1 and u_2 is equal to the similarity of u_3 and u_2, even if the weights used for u_1 and u_2 are smaller. Therefore, computing the user-to-user similarity on highly correlated items (to a target item) or weakly correlated items does not change the final similarity and it remains equal to 1. On the contrary, $WPCC_{no-norm}$ takes into account the absolute values of the weights. This leads to a higher user-to-user correlation if computed on the item ratings which have bigger weights, i.e., are more correlated with the target item. Therefore, the users whose correlation is computed on the items with small weights will likely not to be in the target user neighborhood.

5.2 Evaluating Weighting Schemas

The second experiment evaluates all the weighting schemas described in Subsection 3.1 using $WPCC_{no-norm}$ as the weights incorporation method. In Figure 2 the performance of all these approaches varying the number of nearest neighbors are shown for the two considered data sets. Here, the baseline method uses standard (unweighted) PCC. The performance is shown while varying the number of nearest neighbors.

It cab be seen that SVD-PCC improves MAE and F of the baseline prediction for both data sets and for all the neighborhood sizes. The situation with other weighting schemas is not so clear. For example, IPCC can reduce MAE, especially on the Yahoo! data set, and sometimes outperforms the SVD-PCC schema. The improvements of IPCC over the baseline method are smaller for the MovieLens dataset (considering MAE), however, this is achieved for all the neighborhood sizes. Surprisingly, the situation is different for F measure. While SVD-PCC still outperforms the baseline method, IPCC performs similarly to or even worse than the baseline. This is an interesting observation since IPCC is similar to SVD-PCC; the only difference is that SVD-PCC computes the correlation on a lower dimension item representation rather than on the original raw data.

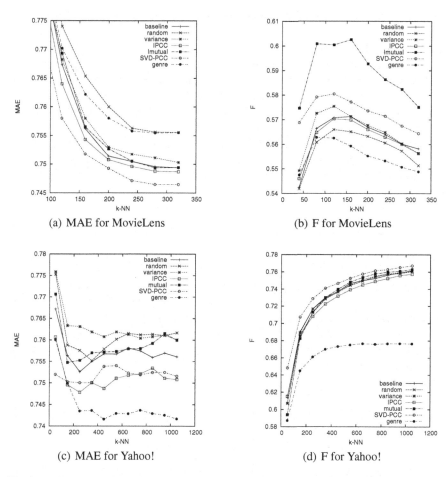

(a) MAE for MovieLens

(b) F for MovieLens

(c) MAE for Yahoo!

(d) F for Yahoo!

Fig. 2 Performance of Item weighting using $WPCC_{no-norm}$ on two data sets

Another interesting behavior can be observed for the weighting schema based on Mutual Information for the MovieLens dataset (called "mutual" in all the Figures). This method performs similarly to the baseline with respect to MAE, however it gets a considerable improvement of F measure. After investigating the method further, we concluded that the increase of the F measure is due to a *large* increase of recall (up to 10%) at the cost of a small decrease in precision. This big improvement of recall is not clear and requires further investigation. Moreover, Mutual Information performs similarly to the baseline method for the Yahoo! data set. This result is different from the one reported in [28] where authors observed a big reduction of the MAE. Note that in [28] authors used EachMovie data set and trained the weights on a small random subset of the users. We guess that the differences could also be explained by the unstable behavior of Mutual Information weighting method. Variance weighting in general does not improve baseline CF, which confirms the results of [12]. This method improves F measure, but it also increases MAE error. As expected, the random weighting schema always decreases the performance of the baseline CF.

Another interesting result can be observed for the genre weighting schema. In the MovieLens dataset it performs even worse than random item weighting and in fact it is the worst performing method. On the contrary, in the Yahoo! data set we have observed a significant reduction of MAE compared to the baseline method. It can be explained by analyzing the way the genre weights are computed. It is often the case that two movies do not share a single genre. Therefore, the genre overlap is 0, making the normalized weights also equal to 0. Hence, when using the genre weighting schema we perform a lot of item filtering together with item weighting.

5.3   Evaluation of Item Filtering

As explained in Section 4, in item filtering we set to zero all weights that are smaller than a given threshold. Our conjecture is that in this way we can reduce MAE by sacrificing the coverage. In Figure 3 the performance of item filtering, with SVD-PCC weighting schema, for all three possible weight incorporation methods is shown. We measure MAE and coverage for different filtering thresholds both for Movie-Lens and Yahoo! datasets. We fixed the neighborhood size for the MovieLens and Yahoo! datasets to be 200 and 300 respectively. For these neighborhood sizes the weighting methods have good MAE without decreasing F measure.

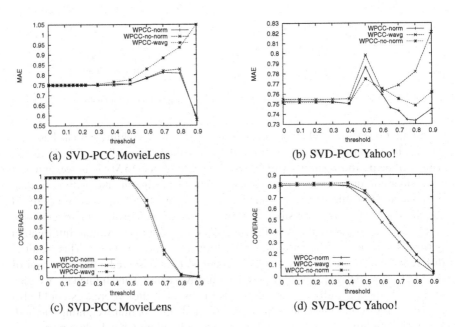

Fig. 3 Performance of item filtering

Note that some changes in the performance of item filtering can be detected only when the applied threshold is larger than 0.3. For the Yahoo! data set the normalized (in [0, 1]) weights are distributed as shown in Figure 4(a). In fact, only 8% of the weights have a value smaller than 0.3 and therefore filtering out these weights has almost no influence on the performance of the algorithms. Conversely, excluding items with weights larger than 0.3 decreases the performance. However, when most of the items are excluded (threshold equal to 0.9 for MovieLens and 0.8 for Yahoo!), the CF algorithms can make a prediction just for a very small number of items (low coverage), and MAE decreases. Hence, using item filtering with the SVD-PCC weighting schema, we can make more accurate predictions only for a very small number of items. In conclusion, using SVD-PCC, one can increase accuracy only by discarding most of the items and scarifying the coverage.

5.4   Increasing the Influence of the Weights

In the previous evaluation of item weighting we used three different methods for weight incorporation into PCC. One of these approaches improved the performance of CF over the baseline method. In the experiments we are illustrating now, we tried to evaluate if amplifying the importance of the weights it is possible to obtain some improvement. Raising each weight to a power higher tan one would make a bigger relative separation between weights. We want to evaluate if this larger separation of the weights can improve the accuracy of the prediction.

The Probability Density Function (PDF) of the weights distribution is shown in Figure 4. Here the weights computed by PCC-SVD for the Yahoo! dataset raised to different powers from 1 to 3 (x axis) are shown. Subfigure 4(a) shows the original (normalized) weight distribution, where most of the weights are distributed in the range [0.2, 0.8]. Subfigure 4(b) shows the PDF of the same weights were each weight was raised to the power of 2. In this case, weights which are close to 1 marginally decrease their value, whereas smaller weights are more strongly pushed towards 0. Therefore, we get a bigger relative separation between large weights (close to 1) and small weights (close to 0). If we raise the weights to a higher power, all the weights

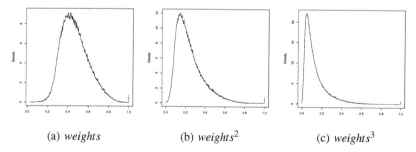

(a) *weights*   (b) *weights^2*   (c) *weights^3*

Fig. 4 Probability density estimation (PDF) of the weight distribution for weight transformations (raise to the power)

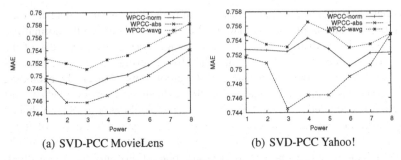

(a) SVD-PCC MovieLens   (b) SVD-PCC Yahoo!

Fig. 5 Performance of SVD-PCC w.r.t. different powers of weights

would move towards zero, except the weights of the items perfectly correlated with the target item, i.e., with value equal to 1.

Figure 5 shows the results of an experiment where we varied the power to which all the weights are raised. Here we used the PCC-SVD weighting schema, and fixed k equal to 200 for the MovieLens dataset and 300 for the Yahoo! dataset. As we expected, raising the weights to a small power (two or three) decreases MAE. For the lack of space we do not show the corresponding results for the F measure, but similarly to results shown before the best performance is obtained raising the weight to the power of 2 and 3.

6   Conclusions

This paper analyzes a wide range of item weighting techniques suitable for CF. Each item weighting technique analyzes the data to get additional statistics about the relationships between the items. This information is then used to fine-tune the user-to-user similarity and to improve the accuracy of the CF recommendation technique. We have observed that the newly-introduced SVD-PCC weighting schema — designed to work for sparse data — performs better than the baseline CF method in the two datasets that was considered, and for all the accuracy measures that we used (MAE, F, Coverage). All the other methods do not have such a stable behavior and they outperform the baseline CF only for some of the error measures, and sometimes they perform even worse. We also experimentally evaluated three different weight incorporation techniques and explained their behavior. In the data sets that we used $WPCC_{no-norm}$ showed a better performance than the other techniques.

This paper gives also an insight into the limitations of the weighting techniques when used for item filtering. The SVD-PCC weighting schema, which has good item weighting performance, showed worse results than the baseline method when applied to item filtering. On the other hand, item filtering with the genre weighting schema resulted in a reduction of MAE at the cost of a reduction of the coverage. Such an approach could be useful for applications where it is important to provide a small number of recommendations. If well used, it could also provide a powerful technique for determining a small number of well targeted recommendations.

In the future, we want to analyze item filtering in depth and understand if it can be used to improve accuracy with other weighting schemas. If successful, this approach could be used in distributed or cross-domain recommender systems, where it is important to understand what portion of items are relevant and should be retrieved from the other domain in order to improve the recommendation [6].

Weight amplification techniques needs some further study. In our last experiment, we tried to re-scale the original weights transforming them with a power function. This is just one possible approach and a better analysis of this issue is required. In fact, the weights computed by our proposed methods provide just a relative measure of the importance of the items, and determining the correct factor that must be used in the PCC metric requires some more experiments.

References

1. Adomavicius, G., Sankaranarayanan, R., Sen, S., Tuzhilin, A.: Incorporating contextual information in recommender systems using a multidimensional approach. ACM Transactions on Information Systems 23(1), 103–145 (2005)
2. Adomavicius, G., Tuzhilin, A.: Toward the next generation of recommender systems: A survey of the state-of-the-art and possible extensions. IEEE Transactions on Knowledge and Data Engineering 17(6), 734–749 (2005)
3. Aha, D.W.: Feature weighting for lazy learning algorithms. In: Feature extraction, construction and selection: a data mining perspective, vol. SECS 453, pp. 13–32. Kluwer Academic, Boston (1998)
4. Baltrunas, L., Ricci, F.: Dynamic item weighting and selection for collaborative filtering. In: Berendt, B., Mladenic, D., Semeraro, G., Spiliopoulou, M., Stumme, G., Svatek, V., Zelezny, F. (eds.) Web Mining 2.0 International Workshop located at the ECML/PKDD 2007, pp. 135–146 (2007)
5. Baltrunas, L., Ricci, F.: Locally adaptive neighborhood selection for collaborative filtering recommendations. In: Nejdl, W., Kay, J., Pu, P., Herder, E. (eds.) AH 2008. LNCS, vol. 5149, pp. 22–31. Springer, Heidelberg (2008)
6. Berkovsky, S., Kuflik, T., Ricci, F.: Cross-domain mediation in collaborative filtering. In: Conati, C., McCoy, K., Paliouras, G. (eds.) UM 2007. LNCS, vol. 4511, pp. 355–359. Springer, Heidelberg (2007)
7. Billsus, D., Pazzani, M.J.: Learning collaborative information filters. In: Shavlik, J.W. (ed.) ICML, pp. 46–54. Morgan Kaufmann, San Francisco (1998)
8. Breese, J.S., Heckerman, D., Kadie, C.M.: Empirical analysis of predictive algorithms for collaborative filtering. In: Cooper, G.F., Moral, S. (eds.) UAI, pp. 43–52. Morgan Kaufmann, San Francisco (1998)
9. Deerwester, S.C., Dumais, S.T., Landauer, T.K., Furnas, G.W., Harshman, R.A.: Indexing by latent semantic analysis. Journal of the American Society of Information Science 41(6), 391–407 (1990)
10. Geng, X., Liu, T.Y., Qin, T., Li, H.: Feature selection for ranking. In: SIGIR 2007: Proceedings of the 30th annual international ACM SIGIR conference on Research and development in information retrieval, pp. 407–414. ACM, New York (2007)
11. Golub, G.H., Van Loan, C.F.: Matrix Computations (Johns Hopkins Studies in Mathematical Sciences). The Johns Hopkins University Press (1996)
12. Herlocker, J.L., Konstan, J.A., Borchers, A., Riedl, J.: An algorithmic framework for performing collaborative filtering. In: SIGIR, pp. 230–237. ACM, New York (1999)

13. Herlocker, J.L., Konstan, J.A., Terveen, L.G., Riedl, J.: Evaluating collaborative filtering recommender systems. ACM Transactions on Information Systems 22, 5–53 (2004)
14. Jin, R., Chai, J.Y., Si, L.: An automatic weighting scheme for collaborative filtering. In: SIGIR 2004: Proceedings of the 27th annual international ACM SIGIR conference on Research and development in information retrieval, pp. 337–344. ACM Press, New York (2004)
15. Kohavi, R., John, G.H.: Wrappers for feature subset selection. Artificial Intelligence 97(1-2), 273–324 (1997)
16. Langley, P.: Selection of relevant features in machine learning. In: Proceedings of the AAAI Fall Symposium on Relevance. AAAI Press, Menlo Park (1994)
17. Mitchell, T.M.: Machine Learning. McGraw-Hill, New York (1997)
18. MovieLens dataset, http://www.grouplens.org/
19. Radlinski, F., Joachims, T.: Query chains: learning to rank from implicit feedback. In: KDD 2005: Proceeding of the eleventh ACM SIGKDD international conference on Knowledge discovery in data mining, pp. 239–248. ACM Press, New York (2005)
20. Resnick, P., Varian, H.R.: Recommender systems. Communications of the ACM 40(3), 56–58 (1997)
21. Salton, G., Mcgill, M.J.: Introduction to Modern Information Retrieval. McGraw-Hill Inc., New York (1986)
22. Sarwar, B., Karypis, G., Konstan, J., Riedl, J.: Application of dimensionality reduction in recommender systems – a case study. In: Proceedings of the WebKDD 2000 Workshop at the ACM-SIGKDD Conference on Knowledge Discovery in Databases (2000)
23. SAS Institute Inc., SAS OnlineDoc(TM), Version 7-1 Cary, SAS Institute Inc., NC (1999)
24. Schafer, J.B., Frankowski, D., Herlocker, J., Sen, S.: Collaborative filtering recommender systems. In: Brusilovsky, P., Kobsa, A., Nejdl, W. (eds.) Adaptive Web 2007. LNCS, vol. 4321, pp. 291–324. Springer, Heidelberg (2007)
25. Doug Rohde's SVD C Library, version 1.34, http://tedlab.mit.edu/~dr/svdlibc/
26. Wettschereck, D., Aha, D.W., Mohri, T.: A review and empirical evaluation of feature weighting methods for a class of lazy learning algorithms. Artif. Intell. Rev. 11(1-5), 273–314 (1997)
27. Yahoo! Research Webscope Movie Data Set. Version1.0, http://research.yahoo.com/
28. Yu, K., Xu, X., Ester, M., Kriegel, H.P.: Feature weighting and instance selection for collaborative filtering: An information-theoretic approach*. Knowledge and Information Systems 5(2), 201–224 (2003)

Using Term-Matching Algorithms for the Annotation of Geo-services

Miha Grčar, Eva Klien, and Blaž Novak

Abstract. This paper presents an approach to automating semantic annotation within service-oriented architectures that provide interfaces to databases of spatial-information objects. The automation of the annotation process facilitates the transition from the current state-of-the-art architectures towards semantically-enabled architectures. We see the annotation process as the task of matching an arbitrary word or term with the most appropriate concept in the domain ontology. The term matching techniques that we present are based on text mining. To determine the similarity between two terms, we first associate a set of documents [that we obtain from a Web search engine] with each term. We then transform the documents into feature vectors and thus transition the similarity assessment into the feature space. After that, we compute the similarity by training a classifier to distinguish between ontology concepts. Apart from text mining approaches, we also present an alternative technique, namely Google Distance, which proves less suitable for our task. The paper also presents the results of an extensive evaluation of the presented term matching methods which shows that these methods work best on synonymous nouns from a specific vocabulary. Furthermore, the fast and simple centroid-based classifier is shown to perform very well for this task. The main contribution of this paper is thus in proposing a term matching algorithm based on text mining and information retrieval. Furthermore, the presented evaluation should give a notion of how the algorithm performs in various scenarios.

Keywords: geo-services, semantic annotation, text mining, search engine querying, machine learning, term matching.

Miha Grčar, Blaž Novak
Jožef Stefan Institute, Jamova 39, 1000 Ljubljana, Slovenia
e-mail: {miha.grcar,blaz.novak}@ijs.si

Eva Klien
Fraunhofer IGD, Fraunhoferstr. 5, 64283 Darmstadt, Germany
e-mail: eva.klien@igd.fraunhofer.de

B. Berendt et al. (Eds.): Knowl. Disc. Enhan. with Sem. and Soc. Info., SCI 220, pp. 127–143.
springerlink.com   © Springer-Verlag Berlin Heidelberg 2009

1   Introduction and Motivation

This paper presents an approach to automating semantic annotation within service-oriented architectures that provide interfaces to databases of spatial-information objects. The automation of the annotation process facilitates the transition from the current state-of-the-art architectures towards semantically-enabled architectures. The techniques presented in this paper are being developed in the course of the European project SWING[1] which deals with introducing semantics into spatial-data infrastructures to support discovery, composition, and execution of geo-services.

In this work, semantic annotation is understood as the process of establishing explicit links between geographic information that is served via OGC [2] services and the vocabulary defined in the domain ontology (i.e. the vocabulary of a specific geo-informatics community). Once the bridge between the two sides is established, the domain ontology can be employed to support all sorts of user tasks.

The main purpose of this paper is to present data mining techniques that facilitate the annotation process. The annotation process can be seen as the task of matching an arbitrary word or term with the most appropriate concept in the domain ontology. Most of the term matching techniques that we present are based on text mining (see Sections 4.1–4.2). To determine the similarity between two terms, we first associate a set of documents with each term. To get the documents, we query search engines, on-line encyclopaedias, dictionaries, thesauri, and so on (query being the term in question). We then transform the documents into feature vectors. By doing so, we transition the similarity assessment into the feature space. Several text mining approaches are at hand to compute the similarity in the feature space either by computing centroids or by training classifiers to distinguish between ontology concepts. Apart from the techniques based on document similarity, we also present Google Distance (see Section 4.3). All the techniques are demonstrated on an example from the domain of mineral resources in Section 5. The text mining techniques which prove to be faster and more accurate are further evaluated in a larger experimental setting presented in Section 6.

2   Related Work

Several knowledge discovery (mostly machine learning) techniques have been employed for ontology learning tasks in the past [6]. Text mining seems to be a popular approach to ontology annotation because the text mining techniques are shown to produce relatively good results. We reference much of the related work from the corresponding sections. In the context of text mining we discuss centroid computation and classification [1] (see Section 4.1.1), k-NN, and classification in general [8]

---

[1] Semantic Web Services Interoperability for Geospatial Decision Making (FP6-026514) http://www.swing-project.org

[2] Open Geospatial Consortium, http://www.opengeospatial.org

(see Section 4.1.2). Apart from the text learning techniques we also employ Google Distance [2] (see Section 4.3).

3   Baseline for the Annotation of Geo-services

Normally, geo-data is served by a database of spatial-information objects through a standardized interface. In our work, we use OGC-defined standard interfaces, namely Web feature services (WFS) [9], to access spatial-information objects. Web feature services are required to implement the capability to describe objects (termed "features") that they serve (see Fig. 1, the left-hand side). These descriptions (schemas) contain the definition of each available class of objects in a similar fashion as a data structure is defined in an object-oriented modeling or programming language: the schema provides the class name and its attributes; each attribute is described by its own name and the corresponding data type.

On the other hand we have real-word entities such as trees, rivers, minerals, quarries, and so on. These are modeled as axiomatized concept definitions in the form of a domain ontology that captures a specific view on the world (see Fig. 1, the right-hand side). The core idea is to employ the domain ontology for the discovery and composition of Web feature services, and also for the retrieval of spatial-information objects (i.e. for the invocation of Web feature services).

In order to support these user tasks, we need to establish a bridge between the WFS schema on one side and the domain ontology on the other. The process of establishing this link is called annotation. We see the annotation as a two-step process. The first step is a simple syntactic translation from a WFS schema description into the appropriate ontology language. We use WSML [3] as the ontology-description language in our work. We call a WFS described in WSML a "feature-type ontology" (FTO). FTOs

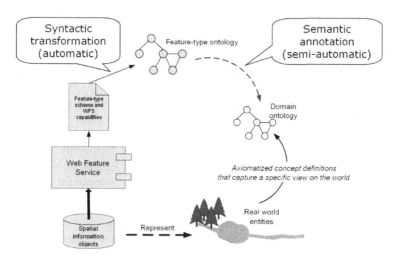

Fig. 1 The two-step semantic annotation process

do not differ much from the original WFS descriptions apart from being formulated in a different description language.

The first step thus establishes the syntactic compatibility between a WFS and the domain ontology (i.e. both descriptions are put into the same ontology-description language). However, the two descriptions are not yet semantically interlinked. The second step thus associates concepts and properties from FTOs with the domain ontology concepts. This process is described in the following section.

4   Automating the Annotation Process

Let us first define the problem of mapping one concept to another in more technical terms. We are given a feature-type ontology (FTO) concept as a single textual string (e.g. OpenPitMine) and a domain ontology which is basically a directed graph in which vertices represent concepts and edges represent relations between concepts. Each concept in the domain ontology is again given as a single textual string (e.g. D:Quarry[3]). The task is now to discover that OpenPitMine is more closely related to D:Quarry as to for instance D:Legislation or D:Transportation. Also important to mention is that every FTO concept has a set of attributes. Each attribute is given as a single textual string (e.g. OpenPitMine.SiteName) and has its corresponding data type (the data type is not expected to provide much guidance in the annotation process since it is usually simply *string*). Concepts in the domain ontology can similarly be described with the surrounding concepts, e.g. D:Quarry-hasLocation-Quarry-Location [4].

A straightforward approach would be to try to compare strings themselves. Even by taking attribute strings into the account coupled with some heuristics we cannot hope for good results this can serve merely as a baseline.

In the following we present several promising approaches that use alternative data sources from the Web to discover mappings between concepts. We limit ourselves to a scenario where attributes are not available (i.e. we are given merely an FTO concept and a set of domain ontology concepts). The task is to arrange domain ontology concepts according to the relatedness to the FTO concept. In the toy examples we will use OpenPitMine as the observed FTO concept and domain ontology concepts D:Quarry, D:Legislation, and D:Transportation.

In Section 4.1 we first introduce the idea of concept comparison by populating concepts with (textual) documents that reflect semantics of these concepts. To enable the realization of these ideas in the context of OGC services we first need to resolve the fact that the concepts are not a-priori populated with documents. Section 4.2 presents two promising techniques of using a Web search engine (in our particular case: Google) to acquire the "missing" documents. The subsequent section (4.3) presents an alternative way of using the Web for the annotation: comparison of

---

[3] With prefix D: we denote concepts that belong to the domain ontology.

[4] This denotes a domain ontology concept named Quarry with "attribute" QuarryLocation which is linked to the concept via the hasLocation relation.

terms by Google Distance. Rather than dealing with documents, this approach deals with term co-occurrences.

4.1 Comparing Documents to Determine Concept Similarity

Suppose we have a set of documents assigned to a concept and that these documents "reflect" the semantics of the concept. This means that the documents are talking about the concept or that the domain expert would use the concept to annotate (categorize) these documents.

In such cases we can compute the similarity between two concepts. We are given an FTO concept (in our case OpenPitMine) and several domain ontology concepts (in our case D:Quarry, D:Transportation, and D:Legislation) with their corresponding document sets (see Fig. 2). We first convert every document into its bag-of-words representation, i.e. into the *tfidf* representation [6]. A *tfidf* representation is actually a sparse vector of word-frequencies (compensated for the commonality of words this is achieved by the *idf* component and normalized). Every component of a *tfidf* vector corresponds to a particular word in the dictionary. With "dictionary" we refer to all the different words extracted from the entire set of documents. If a word does not occur in the document, the corresponding *tfidf* value is missing hence the term

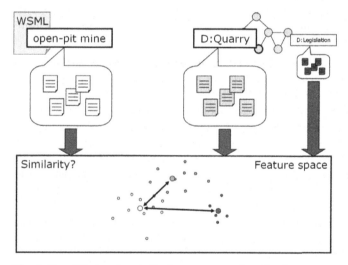

Fig. 2 Transitioning the similarity assessment into the feature space by transforming the documents into the corresponding *tfidf* feature vectors. This figure also illustrates how the comparison of centroids (discussed later on in Section 4.1.1) can be used to conclude that OpenPitMine (represented with white documents/dots) is associated with D:Quarry (represented with light gray documents/dots) stronger than with D:Legislation (represented with dark gray documents/dots). This conclusion is based on the fact that the white centroid is closer to the light gray centroid than to the dark gray centroid. The centroids are represented with the larger dots.

"sparse vector". Each of these vectors belongs to a certain concept we say that the vector is *labelled* with the corresponding concept. This gives us a typical supervised machine learning scenario. In the following subsections we present two different approaches to concept-concept similarity assessment using the centroid-centroid similarity computation approach and the k nearest neighbors classification algorithm.

Comparing Centroids to Determine Concept Similarity. A centroid is a (sparse) vector representing an (artificial) "prototype" document of a document set. Such prototype document should summarize all the documents of a given concept. There are several ways to compute the centroid (given *tfidf*s of all documents in the corresponding set). Some of the well-known methods are the Rocchio formula, average of vector components, and (normalized) sum of vector components. Of all the listed methods, the normalized sum of vector components is shown to perform best in the classification scenario [1]. In the following we limit ourselves to the method of normalized sum. We first represent documents of a particular concept C as normalized *tfidf* vectors $\vec{d_i}$. Now we compute the centroid as given in Eq. 1.

$$\vec{c} = \frac{1}{||\vec{c}||} \sum_{d_i \in C} \vec{d_i} \tag{1}$$

Having centroids computed for all the concepts, we can now measure similarity between centroids and interpret it as similarity between concepts themselves (we are able to do this because a centroid summarizes the concept it belongs to). This is illustrated in Fig. 2. Usually we use cosine similarity measure [6] to measure similarity between two centroid vectors.

Employing Classification to Determine Concept Similarity. We already mentioned that every *tfidf* vector is labelled with the corresponding concept and that this gives us a typical supervised machine learning scenario. In a typical supervised machine learning scenario we are given a set of training examples. A training example is actually a labelled (sparse) vector of numbers. We feed the training examples to a classifier which builds a model. This model summarizes the knowledge required to automatically assign a label (i.e. a concept) to a new yet unlabelled example (we term such unlabelled examples "test examples"). This in effect means that we can assign a new document to one of the concepts. We call such assignment (of a document to a concept) classification.

How do we use classification to compare two concepts? The approach is quite straightforward. We take the documents belonging to a particular FTO concept (in our case the documents of OpenPitMine) and strip them of their label thus forming a test set. Now we assign each of these documents to one of the domain ontology concepts (i.e. we classify each of the documents to one of the domain ontology concepts). The similarity between an FTO concept and a domain ontology concept is simply the number of FTO-concept documents that were assigned to that particular domain ontology concept. Many classifiers assign the same document to all the concepts at the same time but with different probabilities or confidence. We can compute the sum of these probabilities/confidence values instead of simply counting the documents.

There are many different classifiers at hand in the machine learning domain. Herein we discuss a very popular classifier the k-nearest neighbors (k-NN) algorithm which has the property of a "lazy learner". The latter means that k-NN does not build a model out of the training examples instead, it uses them directly to perform classification.

Classification with k-NN. We already mentioned that k-NN is one of the "lazy learners" which means that it does not build a model out of training examples. It performs the classification of a document by finding k most similar documents of all the documents that belong to the domain ontology concepts. The similarity between the document and a domain ontology concept can be computed as the number of documents (from the set of k most similar documents) that belong to that domain ontology concept. Instead of simply counting the documents we can compute the sum of the corresponding cosine similarities. As an alternative to defining a constant neighborhood size, we can set k dynamically by taking all the documents within the cosine similarity range of less than (or equal to) a predefined threshold into account.

4.2   Google Definitions and Contextualized Search Results

"If Google has seen a definition for the word or phrase on the Web, it will retrieve that information and display it at the top of your search results. You can also get a list of definitions by including the special operator define: with no space between it and the term you want defined. For example, the search define:World Wide Web will show you a list of definitions for World Wide Web gathered from various online sources." (excerpt from Google Help Center [5])

Googlebots crawl the Web all the time. In their expeditions they gather terabytes of data which is then processed in order to discover information that is potentially of particular interest to Google users (such as products for sale on-line, weather forecast, travel information, and images). One of such separately maintained information repositories are the definitions of words or phrases as found on the Web.

Google definitions can be used to compensate for the missing document instances each definition (known by Google) can be seen as one document. In this way we can "populate" concepts with documents and then perform the mapping (i.e. the annotation) as already explained in Section 4.1.

To get back to our example, if we populate concepts OpenPitMine, D:Quarry, D:Transportation, and D:Legislation with document instances and then compare OpenPitMine (which is an FTO concept) to the domain ontology concepts (i.e. the other three concepts), we get centroid-to-centroid similarities as shown in Table 1. Since it is hard to find definitions for n-grams such as "open pit mine" (i.e. 3 or more words in a composition), we additionally query Google for the definitions of

---

[5] http://www.google.com/help/features.html#definitions

"pit mine" and "mine", weighting the contribution of these definitions less than the one of the initial composed word (if the "complete" definition exists, that is).

There are still some issues that need to be considered when using this approach. For one, a word can have several meanings, i.e. its semantics depends on the context (or the domain). Google does not know in which context we are searching for a particular definition it thus returns all definitions of a particular word or phrase it keeps in its database. "Mine", for instance, can be defined either as "excavation in the earth from which ores and minerals are extracted" or as "explosive device that explodes on contact". It is important to somehow detect the documents that do not talk about the geospatial domain and exclude them from the annotation process.

Note that we can also populate concepts with Google search results (in contrast or even in addition to populating it with definitions). In this case we can put the search term into a context by extending it with words or phrases describing the context. For example: to populate concept OpenPitMine in the context of "extracting materials" with documents, we would query Google for "open pit mine extracting materials" and consider for instance the first 50 search results. Centroid-to-centroid similarities for this approach are also shown in Table 1.

4.3   Google Distance

Word similarity or word association can be determined out of frequencies of word (co-)occurrences in text corpora. Google Distance [2] uses Google to obtain these frequencies. Based on two requirements, namely (1) if the probability of word w_1 co-occurring with word w_2 is high then the two words are "near" to each other and vice versa, and (2) if any of the two words is not very common in the corpus, the distance is made smaller, the authors came up with the equation given in Eq. 2. They call it Normalized Google Distance (NGD).

$$\text{NGD}(w_1, w_2) = \frac{\max\{\log f(w_1), \log f(w_2)\} - \log(w_1, w_2)}{\log M - \min\{\log f(w_1), \log f(w_2)\}} \tag{2}$$

In Eq. 2, $f(w)$ is the number of search results returned by Google when searching for w (similarly $f(w_1, w_2)$ is the number of search results returned by Google when searching for pages containing both terms), and M is the maximum number of pages that can potentially be retrieved (we believe that posing no constraints on language, domain, file type, and other search parameters Google indexes around 20 billion pages).

It is also possible to put NGD computation into a context. This can be done simply by extending Google queries (the ones that are used to obtain frequencies) with words that form the context. Note that in this case M must be determined as the number of returned search results when searching for the words that form the context. The performance of NGD on our toy example is evident from Table 1.

We believe that NGD is not really the best way to search for synonymy because synonyms generally do not co-occur. It is more a measure of relatedness or

Table 1 The performance of some of the discussed algorithms on our toy example. In the case of the centroid-to-centroid similarity computation, the numbers represent cosine similarity between the terms; in the case of the Normalized Google Distance, the numbers represent the distance measure, and in the case of k-NN, the sum of all the cosine similarities measured between an FTO document and a domain ontology document from the corresponding neighborhood. The results are given for two different contexts: the general context and the context of "extracting material" (note that the contextualization is not applicable if Google definitions are used as the data source). Top 30 search results are considered when querying the search engine. Only English definitions and English search results are considered. In the case of k-NN, k is set dynamically by taking all the documents within the cosine similarity range of less than (or equal to) 0.06 into account. The number that represents the strongest association between the corresponding two terms is emphasized.

		open pit mine		
		general context	"extracting material"	
Centroid Google search	quarry	0.11	0.39	Cosine similarity
	legislation	0.01	0.09	
	transportation	0.02	0.05	
Centroid Google definitions	quarry	0.39	N/A	
	legislation	0.01	N/A	
	transportation	0.05	N/A	
k-NN Google search	quarry	0.61	2.82	Aggregated cosine similarity
	legislation	0	0.43	
	transportation	0	0.10	
k-NN Google definitions	quarry	3.17	N/A	
	legislation	0	N/A	
	transportation	0.35	N/A	
NGD	quarry	0.02	1.92	Distance
	legislation	0.42	3.55	
	transportation	0.50	1.71	

association. Also note that any other search engine that reports the total number of search results can be used instead of Google.

5   A Preliminary Experiment

We tested the presented methods on a dataset from the domain of minerals. We obtained 150 mineral names together with their synonyms[6]. To list just a few: acmite is a synonym for aegirite, diopside is a synonym for alalite, orthite is a synonym for allanite, and so on. The mineral names were perceived as our domain ontology concepts while the synonyms were perceived as the feature-type ontology concepts. For each of the synonyms, the selected algorithms were used to sort the mineral names according to the strength of the association with the synonym in question. We measured the percentage of cases in which the correct mineral name was in the top 1, 3, 5, and 10 names in the sorted list. In other words, we measured the

---

[6] The dataset can be found at http://www. csudh.edu/oliver/chemdata/minsyn.htm

Table 2 The results of the preliminary experiment

				Accuracy [%]			
Algorithm	Data source	Context	Quotes	Top 1	Top 3	Top 5	Top 10
k-NN	Google search	general	no	82.67	90	92	93.33
Centroid	Google search	general	no	78	91.33	92	94
k-NN	Google def.	general	no	70	76	77.33	79.33
Centroid	Google def.	general	no	76	77.33	78.67	79.33
k-NN	Google search	"site: wikipedia.org"	no	43.33	60.67	70	76
k-NN	Google search	"minerals"	no	86.67	97.33	98.67	100
Centroid	Google search	"minerals"	no	91.33	96.67	98.67	99.33
NGD	Google search	general	no	8.67	18	21.33	30.67
NGD	Google search	"minerals"	no	12.67	21.33	29.33	42.67
k-NN	Google search	general	yes	80.67	89.33	91.33	93.33
Centroid	Google search	general	yes	78	91.33	93.33	94.67
k-NN	Google search	"site: wikipedia.org"	yes	27.33	38.67	39.33	42
k-NN	Google search	"minerals"	yes	88	98	99.33	100
Centroid	Google search	"minerals"	yes	93.33	98.67	99.33	100
NGD	Google search	general	yes	16	26	36.67	54.67
NGD	Google search	"minerals"	yes	11.33	22.67	36.67	58

accuracy of each of the algorithms according to the top 1, 3, 5, and 10 suggested mineral names.

We employed 16 algorithms altogether: 7 variants of k-NN, 5 variants of the centroid classifier, and 4 variants of NGD. We varied the context and the data source (either Google definitions or Google search engine). We also varied whether the order of words in a term matters or not (if the order was set to matter then the term was passed to the search engine in quotes). Top 30 search results were considered when querying the search engine. Only English definitions and English search results were considered. In the case of k-NN, k was set dynamically by taking all the documents within the cosine similarity range of less than (or equal to) 0.06 into account. The final outcome of a k-NN algorithm was computed as a sum of all the cosine similarities measured between a synonym document and a mineral name document from the corresponding neighborhood. Table 2 summarizes the results of the experiment. The best performing algorithm is emphasized in the table.

6   Large-Scale Evaluation

In the following section we present a large-scale evaluation of the term matching methods discussed in this paper. In the evaluation we are not using real-life OGC services but rather lexical databases and thesauruses found on the Web. This is due to the fact that we need the datasets to be aligned (i.e. each source term is associated with the corresponding target term). Furthermore, the idea is to evaluate these algorithms in a somewhat broader scope not necessarily limited to the service annotation task.

6.1 Datasets

STINET thesaurus [4]. The Defense Technical Information Centers Scientific and Technical Information Network (STINET) thesaurus provides a hierarchically organised multidisciplinary vocabulary of technical terms. The terms are related to each other with the following relations: "broader term", "narrower term", "used alone for", "use", "used in combination for", and "use in combination". We included "narrower term" and "used alone for" in our experiments. There are about 16,000 terms in the thesaurus, linked with about 15,000 "narrower" and 3,000 "used alone for" links, which we subsampled to 1,000 pairs of terms for each of the two relations. The "narrower" relation indicates that the category described by one term is a subset of the category described by the other term, and "used alone for" indicates a term that can be used to replace the original term. The terms in this thesaurus are mostly phrases pertaining to natural sciences, technology, and military. The fact that the dataset contains a large number of phrases results in an improvement of the accuracy of the evaluated algorithms if quotes are used when querying the search engine. GEMET [7]. GEMET or "General Multilingual Environmental Thesaurus" was developed for the European Topic Centre on Catalogue of Data Sources (ETC/CDS) and the European Environment Agency (EEA) as a general multilingual thesaurus on environmental subjects. It is organised into 30 thematic groups that include topics on natural and human environment, human activities, effects on the environment, and other social aspects related to the environment. It contains about 6,000 terms related among each other with over 5,000 "broader than" and over 2,000 "related to" links for each of the languages. The "related to" relation represents a weaker link than synonymy or "used alone for", which is nicely reflected in the results. We subsampled the dataset to 1,000 English terms for each of the two relations.

Tourism ontology. The tourism ontology is describing various aspects of tourism and commercial services related to it. It consists of 710 concepts linked together with the "is-a" relation and a large corpus of tourism-related Web pages annotated with the concepts. Almost all of the annotated objects are named entities such as place names, people, etc. As can be seen from the results, the inclusion of named entities significantly improves the matching accuracy, while the accuracy of matching just concept descriptors between themselves lies below average. The ontological part of the dataset was used in full in the experiments, while the annotation data was subsampled to 1,000 named entities annotated with about 60 concepts.

WordNet [10]. WordNet is a lexical database for the English language. The smallest entities in the database are the so called "synsets" which are sets of synonymous words. Currently it contains about 115,000 synsets which form over 200,000 word-sense pairs; a word-sense pair represents a word with the corresponding meaning. Synsets are tagged as nouns, verbs, adjectives, or adverbs. Nouns are linked together with hypernymy, hyponymy, holonymy, and meronymy (distinguishing between substance, part-of, and member meronymy); verbs are linked together with hypernymy, troponymy, and entailment; adjectives are linked to related nouns. This is by far the largest dataset we used. We performed over half of the experiments on this data. Almost all of the relations were included in the evaluation and some of them were included two times in order to test the effects of exchanging the left and right-hand side inputs to the algorithms. Each experiment was performed on a sample of 1,000 related words which were selected independently for every test.

Table3 gives the overview of all the experiments. For each of the experiments it lists the corresponding dataset, relationship, direction of the relationship, and an example. The experiment "gemet-bt", for instance, includes the "broader than" relation from the GEMET dataset. The example given in the table is "traffic infrastructure is a broader term than road network". The experiment "wordnet-hypn", on the other hand, includes the hypernymy relation from WordNet. The corresponding example given in the table is "the term imaginary creature is a hypernym of the term monster".

6.2 Experimental Setting

With respect to the preliminary experiments presented in Section 5 larwe did not limit the search to Wikipedia or Google definitions. We also excluded NGD from the evaluation. The main reason is in its non-scalability as it queries the search engine for term co-occurrence frequencies which would result in approximately 1,000,000 (i.e. 1,000 times 1,000) queries for each of the experiments. Text-based classifiers, on the other hand, issue only about 2,000 queries per classifier per experiment.

We avoided specifying search contexts as specifying a context requires some domain-specific knowledge and is therefore human-dependent. We wanted to see what accuracy can be achieved without the human involvement.

We therefore varied the classification algorithm (either k-NN or the centroid classifier) and whether or not to put terms into quotes when querying the search engine. Top 30 search results were considered when querying the search engine. Only English search results were considered. In the case of k-NN, k was set dynamically by taking all the documents within the cosine similarity range of more than (or equal to) 0.06 into account. The final outcome of a k-NN algorithm was computed as the sum of all the cosine similarities measured between a target document and a document from the corresponding neighborhood. The results are presented in the following section.

6.3 Results

The results of the evaluation are given in Fig. 3. Each bar in the chart represents one experiment. An experiment is determined by the corresponding experimental setting (see Section 6.2), the corresponding classification algorithm, and the flag indicating whether the search terms were put into quotes.

The chart bars are sorted according to the experimental setting so that four consecutive bars correspond to one experimental setting. The first two bars in the quadruple represent the accuracy of the centroid classifier while the second two represent the accuracy of k-NN. The first bar in each of the two pairs corresponds

Table 3 The overview of the experiments. V in brackets indicates that the relation links two verbs, N indicates the linkage of two nouns, and I denotes the inversion of the corresponding relation.

Experiment	Dataset	Relation	Direction	Left hand side example	Right hand side example
gemet-bt	GEMET	broader than	↓	road network	traffic infrastructure
gemet-rel	GEMET	related to	—	mineral deposit	mineral resource
stinet-nt	STINET	narrower than	↓	alkali metals	potassium
stinet-uaf	STINET	used alone for	↓	gauss-seidel method	numerical methods and procedures
Tourism-onto	Tourism ontology	is-a	↑	gliding field	sports institution
tourism-annot	Tourism ontology	instance-of	↑	Maastricht	city
wordnet-csv	WordNet	cause (V)	↑	do drugs	trip out
wordnet-entv	WordNet	entails (V)	↑	snore	sleep
wordnet-hypn	WordNet	hypernym (N)	↓	monster	imaginary creature
wordnet-hypv	WordNet	hypernym (V)	↓	Europeanize	modify
wordnet-insn	WordNet	instance-of (N)	↑	Cretaceous period	geological period
wordnet-mmn	WordNet	member meronym (N)	↑	Neptune	solar system
wordnet-mpn	WordNet	part meronym (N)	↑	shuffling	card game
wordnet-msn	WordNet	substance meronym (N)	↑	water	frost snow
wordnet-mmni	WordNet	member meronym (N,I)	↓	Girl Scouts	Girl Scout
wordnet-mpni	WordNet	part meronym (N,I)	↓	pressure feed	oil pump
wordnet-msni	WordNet	substance meronym (N,I)	↓	rum cocktail	rum
wordnet-syn	WordNet	synonym	—	homemaker	housewife

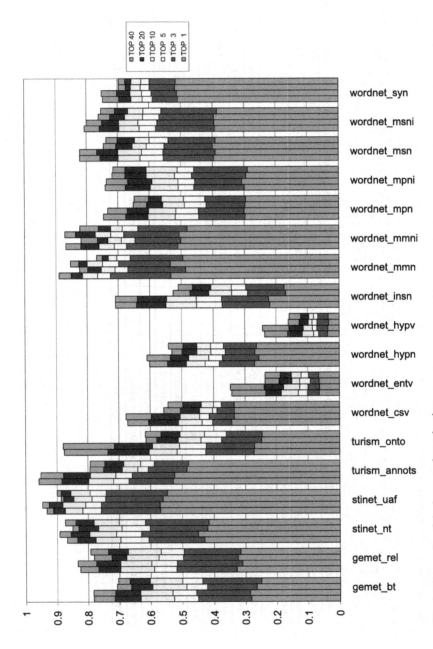

Fig. 3 The results of the large-scale evaluation

to not putting terms into quotes while the second corresponds to putting terms into quotes. The first bar, for instance, corresponds to the "gemet-bt" experimental setting, the centroid classifier was used, and the search terms were not put into quotes. The fourth bar, on the other hand, corresponds to that same experimental setting, but k-NN was used, and the search terms *were* put into quotes. Each bar consists of 6 sections differentiated by different colours corresponding to different accuracy metrics (see the legend in Fig. 3). Since the accuracy of a wider-range metric is always higher than that of a narrower-range metric (e.g. the accuracy at top 10 is always higher than the accuracy at top 5), it is convenient to present the results according to all 6 different metrics in a single bar. Note that the top of the bar section indicates the accuracy according to the corresponding evaluation metric. If we consider the first bar, for example, we can see that when employing the centroid classifier without putting terms into quotes in the "gemet-bt" experimental setting, the correct annotation is automatically determined in roughly 28 % of all the cases. Furthermore, the correct annotation is one of the top 3 suggestions in the sorted list in roughly 45 % of all the cases. If we are willing to consider top 40 suggestions each time, the correct annotation can be determined in roughly 78 % of the cases.

From the evaluation results we can conclude that the terms lexical category (i.e. noun, verb) has the largest impact on the accuracy of the evaluated algorithms. It can be seen that the datasets containing verbs yield the poorest performance by far. This happens due to the fact that the algorithms induce similarities from the contexts defined by the documents corresponding to terms. These contexts are very heterogeneous in the case of a verb.

It is also possible to conclude that the selection of the dataset has a much greater influence on the accuracy of the model than the approach used for constructing it. Within each dataset, the performance of the k-NN classifier is usually slightly worse than that of the centroid-based classifier which is also significantly faster. We believe that this is due to the fact that computing centroids smoothens the noise present in Web search results. k-NN, on the other hand, compares each source-term document to each target-term neighbor document, which can even result in comparing noise to noise.

The use of quotes is beneficial mostly with the datasets that contain a large number of expressions, such as the STINET thesaurus which contains mostly technical expressions. On the other hand, quotes can be detrimental to the accuracy when the terms are already fully determined by a single word and are thus ordinarily referred to by only these words, e.g. "genus Zygnematales" is often referred to as "Zygnematales".

Swapping the left and right-hand side inputs had no significant impact on the accuracy, as can be seen in the WordNets meronymy experiments.

We expected high performance on the datasets where the contexts of matching terms have large overlaps. This is especially true for synonyms and the STINETSs "used alone for" relation. The experiments confirmed our expectations, especially in the case of the "used alone for" relation which yielded a 95 % accuracy at top 40.

Named entities also have very well defined contexts, especially compared to ordinary nouns, therefore datasets that contain many named entities exhibit a better accuracy, as it is the case with the WordNets member meronymy relationship.

7 Conclusions

This paper presents several techniques for automatic annotation in which "external" data sources (such as the Web) are used to compensate for the missing textual documents corresponding to concepts.

According to the preliminary experiment presented in Section 5, the presented techniques have a very good potential. From the results it is evident that at least for the dataset used in the experiment it is not beneficial to limit the search to Wikipedia (a free Web encyclopedia available at http://www.wikipedia.org) or Google definitions. However, it proved useful to perform the search in the context of "minerals". Also important to notice is that k-NN outperforms the centroid classifier when not put into a context. However, when the search is performed in the context of "minerals", the centroid classifier outperforms k-NN. This occurs because the documents, gathered by the contextualized search, are less heterogeneous. Consequently the centroid is able to summarize the topic (which has explicitly to do with "minerals") easier than if the documents were taken from the general context. Even more, the centroid cuts off the outliers (i.e. the noise) by averaging vectors components. On the other hand, when dealing with more heterogeneous set of documents from the general context, the centroid is "shifted" towards irrelevant (sub)topics (i.e. other than "minerals") which results in poorer performance.

Also noticeable from Table 2, as we suspected, NGD is not very successful in detecting synonymy. Furthermore, it is much slower that the other presented algorithms as it is querying the Web to determine term co-occurrences. Last but not least, we can see that the best performing algorithms perform slightly better if the search query is put into quotes (i.e. if the order of words in the term that represents the query is set to matter).

We also performed a large-scale evaluation presented in Section 6. We excluded NGD and querying for Google definitions from these experiments as they proved inferior to their alternatives. From the results in Fig. 3 it is evident that the centroid classifier outperforms k-NN in most of the cases. The accuracy is high for synonymous named entities and nouns, and lower for the other kinds of relations (meronymy, hypernimy). The accuracy is especially low when dealing with verbs instead of nouns (regardless of the relation).

The focus of our future work will be on the implementation of the term matching algorithm based on SVM [11]. We will also consider ways of going beyond string matching by taking the neighborhood of a concept into account similarly to [12] (provided that the concept is a part of an ontology or a similar structure).

The implemented technologies will soon become a part of OntoBridge, an open-source system for semi-automatic data-driven ontology annotation (i.e. for mapping or mediation). OntoBridge will be one of the deliverables of the European project SWING.

Acknowledgments

This work was partially supported by the IST Programme of the European Community under SWING Semantic Web Services Interoperability for Geospatial Decision Making (FP6-026514) and PASCAL Network of Excellence (IST-2002-506778).

References

1. Cardoso-Cachopo, A., Oliveira, L.A.: Empirical Evaluation of Centroid-based Models for Single-label Text Categorization. INSEC-ID Technical Report 7/2006 (2006)
2. Cilibrasi, R., Vitanyi, P.: Automatic Meaning Discovery Using Google (2004)
3. de Bruijn, J.: The Web Service Modeling Language WSML (2005)
4. Public STINET (Scientific & Technical Information Network),
   http://stinet.dtic.mil/str/thesaurus.html
5. Etzioni, O., Cafarella, M., Downey, D., et al.: Web-scale Information Extraction in KnowItAll (Preliminary Results). In: Proceedings of WWW 2004, New York, USA (2004)
6. Grcar, M., Klien, E., Fitzner, D.I., Maué, P., Mladenic, D., Grobelnik, M.: D4.1: Representational Language for Web-service Annotation Models. Project Report FP6-026514 SWING, WP 4, D4.1 (2006)
7. GEMET Thesaurus (General Multilingual Environmental Thesaurus),
   http://www.eionet.europa.eu/gemet/
8. Mitchell, T.M.: Machine Learning. The McGraw-Hill Companies, Inc., New York (1997)
9. Open Geospatial Consortium: Web Feature Service Implementation Specification, Version 1.0.0 (OGC Implementation Specification 02-058) (2002)
10. WordNet, a lexical database of English, http://wordnet.princeton.edu/
11. Vapnik, V.: Statistical Learning Theory. Wiley, New York (1998)
12. Gligorov, R., Aleksovski, Z., ten Warner, K., van Harmelen, F.: Using Google Distance to Weight Approximate Ontology Matches. In: Proceedings of WWW 2007, Banff, Alberta, Canada (2007)

Author Index